Models and Methods for Interval-Valued Cooperative Games in Economic Management

Deng-Feng Li

Models and Methods for Interval-Valued Cooperative Games in Economic Management

Deng-Feng Li
School of Economics and Management
Fuzhou University
Fuzhou, Fujian, China

ISBN 978-3-319-80470-5 ISBN 978-3-319-28998-4 (eBook)
DOI 10.1007/978-3-319-28998-4

Springer Cham Heidelberg New York Dordrecht London

Softcover reprint of the hardcover 1st edition 2016

Printed on acid-free paper

Springer International Publishing AG Switzerland is part of Springer Science+Business Media (www.springer.com)

To my wife, Wei Fei,
and
to my son, Wei-Long Li

Preface

Cooperative games with transferable utility are simply called cooperative games in this book. The cooperative game theory is an important branch of the game theory and has been extensively studied. In (crisp or classical) cooperative games, values (or characteristic functions, payoffs) of coalitions of players are expressed with exact values (i.e., real numbers). However, due to uncertainty and information imprecision in real situations, coalitions' values usually have to be estimated. Recently, intervals are used to estimate inherited imprecision or vagueness in coalitions' values, and hereby there appears an important type of cooperative games with interval data (or interval uncertainty), which often are called interval-valued cooperative games for short. A good example may be the bankruptcy problem with interval data. Interval-valued cooperative games are remarkably different from (classical or crisp) cooperative games since their coalitions' values are expressed with intervals rather than real numbers. Recently, some researchers such as S. Z. Alparslan Gök, R. Branzei, O. Branzei, D. Dimitrov, and S. Tijs paid attention to interval-valued cooperative games and have published some articles. However, most of the existing works used Moore's order relation between intervals or interval arithmetic operations, especially Moore's interval subtraction, which is not invertible and hereby usually enlarges uncertainty of the resulted intervals. This case usually is not accordant with real situations. Thereby, inspired by the companion volume *Linear Programming Models and Methods of Matrix Games with Payoffs of Triangular Fuzzy Numbers* (Deng-Feng Li, 2016, Springer, Heidelberg), in this book, we focus on proposing several commonly used and important interval-valued solution concepts of interval-valued cooperative games and hereby developing some simple, practical, and effective models and methods in which the non-invertible interval subtraction or order relation between intervals is effectively avoided.

This book includes three chapters. Chapter 1 proposes the concept of the interval-valued least square solution of interval-valued cooperative games, establishes quadratic programming models and methods for computing interval-valued least square solutions, and discusses some useful and important properties of interval-valued least square solutions. Chapter 2 studies satisfactory degrees (or ranking indexes) of comparing intervals with the features of inclusion and/or overlap relations and their important properties and proposes the auxiliary satisfactory-degree-based nonlinear programming models for computing interval-valued cores of interval-valued cooperative games and corresponding bisection algorithm. Chapter 3 further expatiates several commonly used and important interval-valued solutions of interval-valued cooperative games and their simplification methods as well as some useful and important properties, including the interval-valued equal division value, the interval-valued equal surplus division value, the interval-valued Shapley value, the interval-valued egalitarian Shapley value, the interval-valued discounted Shapley value, the interval-valued solidarity value, the interval-valued generalized solidarity value, and the interval-valued Banzhaf value. The aim of this book is to develop interval-valued solutions of interval-valued cooperative games and hereby establish their properties, models, methods, and applications, which are remarkably different from the existing studies due to the fact that the non-invertible interval subtraction or order relation between intervals is effectively avoided. I tried my best to ensure that the theoretical models and methods developed in this book are of practicability, simplicity, maneuverability, and universality.

This book is addressed to people in theoretical researches and practical applications from different fields and disciplines such as decision science, game theory, management science, operational research, fuzzy sets or fuzzy mathematics, applied mathematics, industrial engineering, finance, applied economics, expert system, and social economy as well as artificial intelligence. Moreover, it is also addressed to teachers, postgraduates, and doctors in colleges and universities in different disciplines or majors: decision analysis, management, operation research, fuzzy mathematics, fuzzy system analysis, applied mathematics, systems engineering, project management, supply chain management, industrial engineering, applied economics, and hydrology and water resources.

First of all, special thanks are due to my doctoral graduates Jia-Cai Liu and Fang-Xuan Hong for completing and publishing several articles. This book was supported by the Key Program of the National Natural Science Foundation of China (No. 71231003), the National Natural Science Foundation of China (Nos. 71171055, 71461005, 71561008, and 71101033), the "Chang-Jiang Scholars" Program (the Ministry of Education of China), the Program for New Century Excellent Talents in University (the Ministry of Education of China, NCET-10-0020), and the Specialized Research Fund for the Doctoral Program of Higher Education of China (No. 20113514110009) as well as "Science and Technology Innovation Team Cultivation Plan of Colleges and Universities in Fujian Province." I would like to acknowledge the encouragement and support of my wife as well as the understanding of my son.

Last but not least, I would like to acknowledge the encouragement and support of all my friends and colleagues.

Ultimately, I should claim that I am fully responsible for all errors and omissions in this book.

Fuzhou, Fujian, China
October 22, 2015

Deng-Feng Li

Abstract

The focus of this book is to propose several commonly used and important interval-valued solution concepts of interval-valued cooperative games with transferable utility which are called interval-valued cooperative games for short and hereby develop simple and effective models and methods in which the non-invertible interval subtraction or interval order relation is effectively avoided. This book includes three chapters. Chapter 1 proposes the interval-valued least square solutions of interval-valued cooperative games and quadratic programming models and methods as well as properties. Chapter 2 proposes the satisfactory-degree-based nonlinear programming models for computing interval-valued cores of interval-valued cooperative games and corresponding bisection algorithm. Chapter 3 expatiates several interval-valued solutions of interval-valued cooperative games and their simplification methods as well as properties, including the interval-valued equal division and equal surplus division values; the interval-valued Shapley, egalitarian Shapley, and discounted Shapley values; the interval-valued solidarity and generalized solidarity values; and the interval-valued Banzhaf value. The aim of this book is to develop interval-valued solutions of interval-valued cooperative games and hereby establish their properties, models, methods, and applications.

This book is addressed to people in theoretical researches and practical applications from different fields and disciplines such as decision science, game theory, management science, operational research, fuzzy sets or fuzzy mathematics, applied mathematics, industrial engineering, finance, applied economics, expert system, and social economy as well as artificial intelligence. Moreover, it is also addressed to teachers, postgraduates, and doctors in colleges and universities in different disciplines or majors: decision analysis, management, operation research, fuzzy mathematics, fuzzy system analysis, applied mathematics, systems engineering, project management, supply chain management, industrial engineering, applied economics, and hydrology and water resources.

Contents

About the Author

Deng-Feng Li was born in 1965. He received his B.Sc. and M.Sc. degrees in applied mathematics from the National University of Defense Technology, Changsha, China, in 1987 and 1990, respectively, and a Ph.D. degree in system science and optimization from the Dalian University of Technology, Dalian, China, in 1995. From 2003 to 2004, he was a visiting scholar with the School of Management, University of Manchester Institute of Science and Technology, Manchester, UK. He is currently a distinguished professor of "Chang-Jiang Scholars" Program, Ministry of Education of China, and "Min-Jiang Scholarship" distinguished professor with the School of Economics and Management, Fuzhou University, Fuzhou, China. He has been conferred the Outstanding Contribution Experts of the National Middle-Aged and Young of China and was approved as an expert of the Enjoyment of the State Council Special Allowance of China. He has authored or coauthored more than 300 journal papers and 7 monographs. He has coedited 1 proceeding of the international conference and 2 special issues of journals and won 25 academic achievements and awards such as the Chinese State Natural Science Award and the 2013 I.E. Computational Intelligence Society IEEE Transactions on Fuzzy Systems Outstanding Paper Award. His current research interests include classical and fuzzy game theory, fuzzy decision analysis, group decision-making, supply chain management, fuzzy sets and system analysis, fuzzy optimization, and differential game. He is the editor in chief of *International Journal of Fuzzy System Applications* and associate editor and/or editor of several international journals.

Chapter 1
The Interval-Valued Least Square Solutions of Interval-Valued Cooperative Games

Abstract The aim of this chapter is to propose the concept of the interval-valued least square solution of interval-valued cooperative games and develop fast and effective quadratic programming methods for computing such interval-valued least square solutions. In this chapter, after briefly reviewing concepts of solutions of cooperative games and intervals as well as interval operations, based on the least square method and distance measure between intervals, we construct two quadratic programming models and obtain their analytical solutions, which are used to determine players' interval-valued imputations. Hereby the interval-valued least square solutions of interval-valued cooperative games are determined in the sense of minimizing the loss functions. The quadratic programming models and method proposed in this chapter are compared with other methods to show the validity, the applicability, and the advantages.

Keywords Interval-valued cooperative game • Least square method • Loss function • Mathematical programming • Interval computation

1.1 Introduction

Game theory is engaged in competing and strategic interaction among players or subjects in management, economics, finance, business, environment, and engineering [1, 2]. It has gradually developed and formed into two main branches: cooperative games [1, 3] and noncooperative games [4, 5]. There are numerous important works on the theory and applications of noncooperative games [5]. Therefore, our interests are constrained to cooperative games with transferable utility, especially an important kind of cooperative games under uncertain environments or interval data [6, 7]. In this book, all cooperative games are referred to cooperative games with transferable utility unless otherwise stated.

Cooperative games have been extensively studied [8, 9]. In (crisp or classical) cooperative games, values (or characteristic functions, payoffs) of coalitions of players are expressed with exact values (i.e., real numbers) [10, 11]. However, due to uncertainty and information imprecision in real situations, coalitions' values usually have to be estimated. Recently, intervals are used to estimate inherited imprecision or vagueness in coalitions' values and hereby there appears an

D.-F. Li, *Models and Methods for Interval-Valued Cooperative Games in Economic Management*, DOI 10.1007/978-3-319-28998-4_1

important type of cooperative games with interval-valued data, which often are called interval-valued cooperative games for short [7, 12, 13]. A good example may be the bankruptcy problem with interval-valued data [6, 7]. In this example, a certain amount of money (i.e., estate) has to be divided among some people (i.e., claimants) who have individual claims on the estate. In reality, this estate is a positive real number known with certainty whereas the claimants may only give the smallest and the biggest values of the claims, i.e., the claims belong to closed and bounded intervals of positive real numbers. This kind of bankruptcy problems may be correspondingly modeled as interval-valued cooperative games [6, 14]. Interval-valued cooperative games are remarkably different from (classical or crisp) cooperative games since their coalitions' values are expressed with intervals rather than real numbers. Furthermore, no probabilistic assumptions about the range of coalitions' values (i.e., intervals) are known a priori, as is usually the case in real-life economical management practice. Interval-valued cooperative games seem to be suitable for modeling all the economical management situations where the players (or participants) consider cooperation and know with certainty only lower and upper bounds of all potential values/payoffs (or revenues, costs) generated via cooperation.

Recently, interval-valued cooperative games have attracted attention of researchers. Branzei et al. [7] gave a good survey which overviewed and updated the results on interval-valued cooperative games and discussed a variety of existing and potential applications of interval-valued cooperative games in economic management situations where probability distribution is unknown a priori. To be more concrete, Branzei et al. [6] firstly introduced interval-valued cooperative games to handle bankruptcy problems with the numerical estate and interval-valued claims. In a similar way to the Shapley value [10], they proposed two interval-valued Shapley-like values and discussed their interrelations through using the arithmetic of intervals [15]. Alparslan Gök et al. [16] considered selections of interval-valued cooperative games. Such selections are essentially (classical or crisp) cooperative games. In the same way to the definitions of the core [17, 18] and the Shapley value [10] for cooperative games, they straightforwardly defined the interval-valued core and the interval-valued Shapley value of interval-valued cooperative games based on selections' solutions. For instance, the interval-valued core of an interval-valued cooperative game is defined as the union of the cores of all its selections, i.e., the union of the cores of the selected (classical or crisp) cooperative games. Alparslan Gök et al. [12] gave an axiomatic characterization of the interval-valued Shapley-like value on a special subclass of interval-valued cooperative games in which the interval-valued cooperative games are of the so-called size monotonicity. Han et al. [19] introduced the notions of interval-valued cores and the interval-valued Shapley-like value for interval-valued cooperative games according to the Moore's subtraction [15] and the newly defined order relation between intervals. Mallozzi et al. [13] introduced the concept of a core-like for cooperative games with coalitions' values represented by fuzzy intervals (i.e., fuzzy numbers) [20] and a balanced-like condition which is proven to be necessary but not sufficient to guarantee the non-empty of the core-like. Branzei et al. [21] defined the interval-valued core of interval-valued cooperative games by discussing the interval-valued square dominance and interval-valued dominance imputations. Alparslan Gök et al. [22] introduced some set-valued solution concepts of interval-valued cooperative games, which include the interval-valued core, the interval-valued dominance core, and the interval-valued stable sets. However, most of the aforementioned works used the Moore's interval operations [15], especially the Moore's interval subtraction, which usually enlarges uncertainty of the resulted interval. This case usually is not accordant with real situations. In this chapter, we focus on developing simple and effective quadratic programming methods for solving interval-valued cooperative games. More precisely, using the least square method and the concepts of loss functions and distance measures between intervals, we construct two quadratic programming models and obtain their analytical solutions, which are used to determine players' interval-valued imputations. Hereby, the interval-valued least square solutions of interval-valued cooperative games are determined in the sense of minimizing the loss functions. The quadratic programming methods proposed in this chapter are remarkably different from the aforementioned methods. On the one hand, the developed methods can provide analytical formulae for determining the interval-valued least square solutions of interval-valued cooperative games. On the other hand, the developed methods can effectively avoid the Moore's interval subtraction.

The rest of this chapter is organized as follows. In the next section, we briefly review the concepts of solutions of cooperative games. Section 1.3 briefly reviews the concepts of intervals and distances as well as their arithmetic operations and hereby introduces interval-valued cooperative games and defines the loss function to measure differences between interval-valued payoff vectors and interval-type values of players' coalitions. In Sect. 1.4, two quadratic programming models are constructed to compute the interval-valued least square solutions of interval-valued cooperative games. Moreover, some important properties of the interval-valued least square solutions of interval-valued cooperative games are discussed. In Sect. 1.5, the quadratic programming models and methods are illustrated with numerical examples about the optimal allocation of companies' cooperative profits and compared with other similar methods.

1.2 Cooperative Games and Their Solutions

To facilitate the context, in what follows, we briefly review some concepts and notations of cooperative games.

Let $N=\{1,2,\ldots,n\}$ be the set of players i $(i=1,2,\ldots,n)$, where n is a positive integer, and $n\geq 2$. Any subset S of the set N, i.e., $S\subseteq N$, is called a coalition. N is referred to as the grand coalition. $\varnothing$ is called an empty coalition, i.e., an empty set of players. Usually, we denote the set of coalitions of players in the set N by 2^N.

Denote the set of real numbers by R. A n-person cooperative game is an ordered-pair $< N, \upsilon >$, where $\upsilon : 2^N \mapsto R$ is the characteristic function which assigns a value $\upsilon(S) \in R$ to the coalition $S \in 2^N$, and $\upsilon(\varnothing) = 0$. $\upsilon(S)$ is called the value of the coalition S. It can be interpreted as the maximal worth (or profit, reward, cost savings) that the players of the coalition S can obtain when they cooperate. In the sequent, we identify the n-person cooperative game $< N, \upsilon >$ with its characteristic function υ. That is to say, the n-person cooperative game $< N, \upsilon >$ usually is referred to as the cooperative game υ for short. The set of n-person cooperative games is denoted by G^n.

For any coalitions $S \subseteq N$ and $T \subseteq N$ ($S \cap T = \varnothing$), if

$$\upsilon(S \cup T) \geq \upsilon(S) + \upsilon(T),$$

then the cooperative game $\upsilon \in G^n$ is superadditive.

The superadditivity plays an important role in solutions of cooperative games. It means that the greater the coalition the more the value of the coalition. However, large coalitions may be inefficient in that it is more difficult for them to reach agreements on the distribution of their rewards (or profits, values).

The following weak version of the superadditivity is very useful.

For any coalition $S \subseteq N$ and $i \notin S$, if

$$\upsilon(S \cup i) \geq \upsilon(S) + \upsilon(i),$$

then the cooperative game $\upsilon \in G^n$ is weakly superadditive, where $\upsilon(S \cup i) = \upsilon(S \cup \{i\})$ and $\upsilon(i) = \upsilon(\{i\})$. In the sequent, to be more concise, we usually write $\upsilon(S \cup i)$, $\upsilon(S \backslash i)$, $\upsilon(i)$, and $\upsilon(i,j)$ instead of $\upsilon(S \cup \{i\})$, $\upsilon(S \backslash \{i\})$, $\upsilon(\{i\})$, and $\upsilon(\{i,j\})$, respectively.

A more general version of the superadditivity is the convexity, which plays an important role in economical management applications of cooperative games.

For any coalitions $S \subseteq N$ and $T \subseteq N$, if

$$\upsilon(S \cup T) + \upsilon(S \cap T) \geq \upsilon(S) + \upsilon(T),$$

then the cooperative game $\upsilon \in G^n$ is convex.

Obviously, a convex cooperative game $\upsilon \in G^n$ is superadditive. Moreover, it is not difficult to prove that a cooperative game $\upsilon \in G^n$ is convex if and only if for any player $i \in N$ and coalitions $S \subseteq T \subseteq N \backslash i$,

$$\upsilon(S \cup i) - \upsilon(S) \leq \upsilon(T \cup i) - \upsilon(T). \tag{1.1}$$

Thus, the cooperative game is convex if and only if the marginal contribution of any player to each coalition is monotonic and non-decreasing with respect to the set-theoretic inclusion. This well explains the term "convex."

Inspired by Eq. (1.1), we can further define the monotonicity of a cooperative game. Specifically, for any coalitions $S \subseteq N$ and $T \subseteq N$, if $S \subseteq T$ so that

$$v(S) \leq v(T),$$

then the cooperative game $v \in G^n$ is monotonic.

For any coalition $S \subseteq N$, if

$$v(S) + v(N \backslash S) = v(N),$$

then the cooperative game $v \in G^n$ is constant-sum.

As pointed out by Peleg and Sudhölter [23], constant-sum cooperative games have been extensively investigated in the early work in game theory [8]. In addition, very often political cooperative games are constant-sum.

If a cooperative game $v \in G^n$ is additive, i.e., for any coalition $S \subseteq N$,

$$v(S) = \sum_{i \in S} v(i),$$

then the cooperative game v is inessential. Conversely, that is, if there exists a coalition $S \subseteq N$ so that

$$v(S) > \sum_{i \in S} v(i),$$

then the cooperative game $v \in G^n$ is essential.

Obviously, an inessential cooperative game is trivial from a game-theoretic point of view. In other words, if every player $i \in N$ demands at least $v(i)$, then the allocation (or distribution) of $v(N)$ can be uniquely determined.

Let $x_i(v) \in R$ be a payoff (or value) which is allocated to the player $i \in N$ when he/she participates in the cooperative game $v \in G^n$ under the condition that the grand coalition N is reached. Then, $\boldsymbol{x}(v) = (x_1(v), x_2(v), \ldots, x_n(v))^{\mathrm{T}}$ is a payoff vector of n players, where the symbol "T" is a transpose of a vector or matrix. If a payoff vector $\boldsymbol{x}(v)$ satisfies both the efficiency and individual rationality conditions as follows:

$$\sum_{i=1}^{n} x_i(v) = v(N)$$

and

$$x_i(v) \geq v(i) \quad (i = 1, 2, \ldots, n),$$

then the payoff vector $\boldsymbol{x}(v)$ is called an imputation of the cooperative game $v \in G^n$. The set of imputations of a cooperative game $v \in G^n$ is denoted by $I(v)$.

It is obvious that $I(v)$ is empty if and only if $v(N) < \sum_{i=1}^{n} v(i)$. Furthermore, for an inessential cooperative game $v \in G^n$, there exists a unique imputation

$x(v) = (v(1), v(2), \ldots, v(n))^{\mathrm{T}} \in I(v)$. Therefore, in this book, our main interest is essential cooperative games.

Generally, the imputation set of an essential cooperative game is infinite. Consequently, there is a need for some criteria to single out those imputations that are most likely to occur [24]. In this way, we can obtain some subset of the imputation set $I(v)$ as a solution of the essential cooperative game $v \in G^n$. For conciseness, in the sequent, essential cooperative games simply are called cooperative games unless otherwise stated.

In the following, we mainly review two important solutions of cooperative games: the core firstly proposed by Gillies [17] and the Shapley value developed by Shapley himself [10].

Let a core of any cooperative game $v \in G^n$ be denoted by $C(v)$, which is defined as follows:

$$C(v) = \left\{ x(v) \in I(v) \Big| \sum_{i \in S} x_i(v) \geq v(S) \text{ for all } S \subset N \right\}.$$

Clearly, for an inessential cooperative game $v \in G^n$, there exists a unique element in the core $C(v)$, i.e., $C(v) = \left\{ (v(1), v(2), \ldots, v(n))^{\mathrm{T}} \right\} = I(v)$.

If there exists an imputation $x(v) \in C(v)$, then there is no coalition $S \subseteq N$ which has an incentive to split off if $x(v)$ is the proposed reward (or profit) allocation alternative for the grand coalition N. The reason is that the total amount $\sum_{i \in S} x_i(v)$ allocated to the coalition S is not smaller than the amount $v(S)$ which the players can obtain by forming the subcoalition.

In some situations, the core of a cooperative game may be empty. Such an example was given by Owen [1]. In reality, for many cooperative games, their cores are non-empty and include numerous elements. In this case, these cores are polytopes. Then, from the above definition of a core, the core $C(v)$ of a cooperative game $v \in G^n$ can easily be obtained because $C(v)$ is defined with the aid of a finite system of linear inequalities. More precisely, the core $C(v)$ of a cooperative game $v \in G^n$ can be obtained through solving the system of linear inequalities as follows:

$$\begin{cases} \sum_{i \in S} x_i(v) \geq v(S) \text{ for all } S \subset N \\ \sum_{i=1}^{n} x_i(v) = v(N). \end{cases} \tag{1.2}$$

Clearly, the core of cooperative games is a set-valued solution concept, which is more difficult to be applied than a single-valued solution concept.

The Shapley value is an important single-valued solution concept of cooperative games. For an arbitrary cooperative game $v \in G^n$, the Shapley value is defined as a payoff vector $\boldsymbol{\Phi}^{\mathrm{SH}}(v) = \left(\phi_1^{\mathrm{SH}}(v), \phi_2^{\mathrm{SH}}(v), \ldots, \phi_n^{\mathrm{SH}}(v)\right)^{\mathrm{T}}$, whose components are given as follows:

$$\phi_i^{\mathrm{SH}}(v) = \sum_{S \subseteq N \backslash i} \frac{s!(n-s-1)!}{n!}(v(S \cup i) - v(S)) \quad (i = 1, 2, \ldots, n), \tag{1.3}$$

respectively, where s is the cardinality of a coalition $S \subseteq N$, i.e., $s = |S|$. $n!$ is the factorial of n, i.e., $n! = n \times (n-1) \times \cdots \times 2 \times 1$.

For some interpretation of the Shapley value, the reader may further be referred to [1, 10].

1.3 Interval-Valued Cooperative Games and Their Interval-Valued Least Square Solutions

To introduce the concept of interval-valued cooperative games, we firstly review the concepts of intervals and their distances as well as interval arithmetic operations.

1.3.1 Interval Operations and Distances Between Intervals

Denote $\overline{a} = [a_L, a_R] = \{a | a \in R, a_L \leq a \leq a_R\}$, which is called an interval, where R is the set of real numbers stated as the above. $a_L \in R$ and $a_R \in R$ are called the lower bound and the upper bound of the interval $\overline{a}$, respectively. Let $\overline{R}$ be the set of intervals on the set R.

Obviously, if $a_L = a_R$, then the interval $\overline{a} = [a_L, a_R]$ degenerates to a real number, denoted by a, where $a = a_L = a_R$. Conversely, a real number a may be written as an interval $\overline{a} = [a, a]$. Therefore, intervals are a generalization of real numbers. That is to say, real numbers are a special case of intervals [15, 25].

If $a_L \geq 0$, then $\overline{a} = [a_L, a_R]$ is called a non-negative interval, denoted by $\overline{a} \geq 0$. Likewise, if $a_R \leq 0$, then $\overline{a}$ is called a non-positive interval, denoted by $\overline{a} \leq 0$. If $a_L > 0$, then $\overline{a}$ is called a positive interval, denoted by $\overline{a} > 0$. If $a_R < 0$, then $\overline{a}$ is called a negative interval, denoted by $\overline{a} < 0$.

In the following, we give some arithmetic operations of intervals such as the equality, the addition, and the scalar multiplication as follows [15, 25].

Definition 1.1 Let $\overline{a} = [a_L, a_R]$ and $\overline{b} = [b_L, b_R]$ be two intervals on the set $\overline{R}$. The interval arithmetic operations are stipulated as follows:

1. Equality of two intervals: $\overline{a} = \overline{b}$ if and only if $a_L = b_L$ and $a_R = b_R$
2. Addition (or sum) of two intervals: $\overline{a} + \overline{b} = [a_L + b_L, a_R + b_R]$

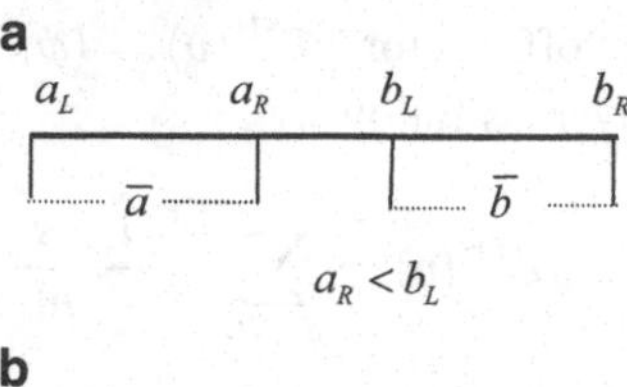

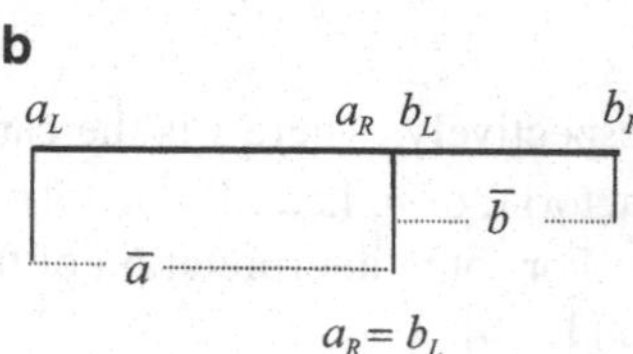

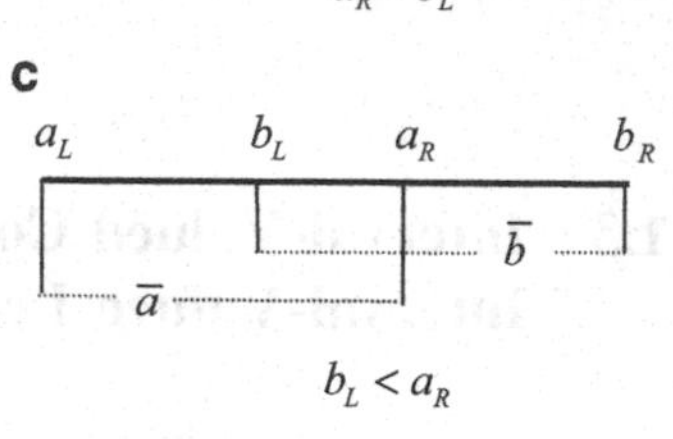

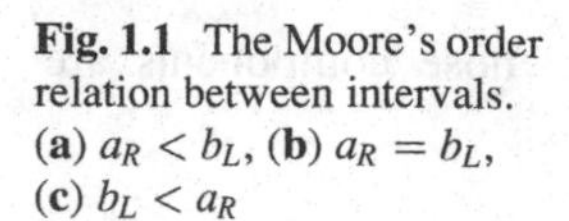
Fig. 1.1 The Moore's order relation between intervals. (**a**) $a_R < b_L$, (**b**) $a_R = b_L$, (**c**) $b_L < a_R$

3. Scalar multiplication of a real number and an interval:

$$\gamma\bar{a} = \begin{cases} [\gamma a_L, \gamma a_R] & \text{if } \gamma \geq 0 \\ [\gamma a_R, \gamma a_L] & \text{if } \gamma < 0, \end{cases}$$

where $\gamma \in R$ is any real number.

Clearly, the above arithmetic operations of intervals are a generalization of those of real numbers.

In real economic management, we usually have to compare or rank intervals, which is a difficult and an important problem. In a similar way to comparison of the real numbers, Moore [15] defined the order relation between intervals as follows:

$$\bar{a} \leq \bar{b} \text{ if and only if } a_L \leq b_L \text{ and } a_R \leq b_R, \tag{1.4}$$

which is simply called the Moore's order relation between intervals, depicted as in Fig. 1.1.

The concept of distances is defined to measure differences between intervals.

Definition 1.2 Let $\bar{a}$ and $\bar{b}$ be two intervals on the set $\bar{R}$. If a mapping $d : \bar{R} \times \bar{R} \mapsto R$ satisfies the three properties 1–3 as follows:

1. Non-negativity: $d(\bar{a}, \bar{b}) \geq 0$
2. Symmetry: $d(\bar{a}, \bar{b}) = d(\bar{b}, \bar{a})$
3. Trigonometrical inequality relation: $d(\bar{a}, \bar{b}) \leq d(\bar{a}, \bar{c}) + d(\bar{c}, \bar{b})$ for any interval $\bar{c}$ on the set $\bar{R}$

then $d(\bar{a}, \bar{b})$ is called the distance between the intervals $\bar{a}$ and $\bar{b}$.

Obviously, Definition 1.2 is a natural generalization of that of the set of real numbers.

Naturally, there are various forms of distances between intervals. For example, to meet the need of modeling interval-valued cooperative games in the subsequent Sect. 1.4, we define the distance between two intervals $\overline{a} \in \overline{R}$ and $\overline{b} \in \overline{R}$ as follows:

$$D(\overline{a}, \overline{b}) = (a_L - b_L)^2 + (a_R - b_R)^2. \tag{1.5}$$

To be more precise, Eq. (1.5) is the square of the distance between the intervals $\overline{a}$ and $\overline{b}$.

Clearly, Eq. (1.5) is very similar to the distance between two points in the two-dimensional Euclidean space R^2.

Theorem 1.1 *$D(\overline{a}, \overline{b})$ defined by Eq. (1.5) is the distance between the intervals $\overline{a} \in \overline{R}$ and $\overline{b} \in \overline{R}$.*

Proof We need to validate that $D(\overline{a}, \overline{b})$ defined by Eq. (1.5) satisfies the three properties 1–3 of Definition 1.2, respectively. It is easy to see from Eq. (1.5) that $D(\overline{a}, \overline{b}) \geq 0$ and $D(\overline{a}, \overline{b}) = D(\overline{b}, \overline{a})$ for any intervals $\overline{a}$ and $\overline{b}$. Namely, $D(\overline{a}, \overline{b})$ satisfies the properties 1 and 2 of Definition 1.2.

For any interval $\overline{c}$ on the set $\overline{R}$, where $\overline{c} = [c_L, c_R]$, it directly follows from Eq. (1.5) that

$$\begin{aligned} D(\overline{a}, \overline{b}) &= (a_L - b_L)^2 + (a_R - b_R)^2 \\ &\leq \left[(a_L - c_L)^2 + (c_L - b_L)^2\right] + \left[(a_R - c_R)^2 + (c_R - b_R)^2\right] \\ &= \left[(a_L - c_L)^2 + (a_R - c_R)^2\right] + \left[(c_L - b_L)^2 + (c_R - b_R)^2\right] \\ &= D(\overline{a}, \overline{c}) + D(\overline{c}, \overline{b}), \end{aligned}$$

i.e.,

$$D(\overline{a}, \overline{b}) \leq D(\overline{a}, \overline{c}) + D(\overline{c}, \overline{b}).$$

Hence, $D(\overline{a}, \overline{b})$ satisfies the property 3 of Definition 1.2. Therefore, we have proven that $D(\overline{a}, \overline{b})$ defined by Eq. (1.5) is the distance between the intervals $\overline{a}$ and $\overline{b}$.

Note that the square appears in Eq. (1.5), which is also the distance from Theorem 1.1. In the sequent, the distance between two intervals is referred to the square of the distance given by Eq. (1.5) unless otherwise specified.

1.3.2 Interval-Valued Cooperative Games and the Interval-Valued Least Square Solutions

A n-person interval-valued cooperative game $\overline{v}$ is an ordered-pair $< N, \overline{v} >$, where $N = \{1, 2, \ldots, n\}$ is the set of players and $\overline{v}$ is the interval-valued characteristic function of coalitions of players, and $\overline{v}(\varnothing) = [0, 0]$. Note that usually $\overline{v}(\varnothing)$ is simply written as $\overline{v}(\varnothing) = 0$ according to the notation of intervals. Stated as earlier, $\varnothing$ is an empty set. Generally, for any coalition $S \subseteq N, \overline{v}(S)$ is denoted by the interval $\overline{v}(S) = [v_L(S), v_R(S)]$, where $v_L(S) \leq v_R(S)$. Stated as earlier, we usually write $\overline{v}(S \backslash i)$, $\overline{v}(S \cup i)$, $\overline{v}(i)$, and $\overline{v}(i,j)$ instead of $\overline{v}(S \backslash \{i\})$, $\overline{v}(S \cup \{i\})$, $\overline{v}(\{i\})$, and $\overline{v}(\{i,j\})$, respectively. In the sequent, a n-person interval-valued cooperative game $< N, \overline{v} >$ is simply called the interval-valued cooperative game $\overline{v}$. The set of n-person interval-valued cooperative games $\overline{v}$ is denoted by $\overline{G}^n$.

Likewise, the superadditivity and the convexity play an important role in solutions of interval-valued cooperative games. Specifically, for any coalitions $S \subseteq N$ and $T \subseteq N$ $(S \cap T = \varnothing)$, if

$$\overline{v}(S \cup T) \geq \overline{v}(S) + \overline{v}(T),$$

then the interval-valued cooperative game $\overline{v} \in \overline{G}^n$ is superadditive.

Analogously, for any coalition $S \subseteq N$ and $i \notin S$, if

$$\overline{v}(S \cup i) \geq \overline{v}(S) + \overline{v}(i),$$

then the interval-valued cooperative game $\overline{v} \in \overline{G}^n$ is weakly superadditive.

Certainly, the weak superadditivity is weaker than the superadditivity. However, the superadditivity is weaker than the convexity. An interval-valued cooperative game $\overline{v} \in \overline{G}^n$ is convex if

$$\overline{v}(S \cup T) + \overline{v}(S \cap T) \geq \overline{v}(S) + \overline{v}(T)$$

for any coalitions $S \subseteq N$ and $T \subseteq N$.

For any positive real number β and an interval-valued cooperative game $\overline{v} \in \overline{G}^n$, according to the case 3 of Definition 1.1, $\beta\overline{v}$ is defined as an interval-valued cooperative game with the interval-valued characteristic function $\beta\overline{v}$, where

$$\beta\overline{v}(S) = \beta[v_L(S), v_R(S)]$$

for any coalition $S \subseteq N$, i.e.,

$$\beta\overline{v}(S) = [\beta v_L(S), \beta v_R(S)]. \tag{1.6}$$

Usually, $\beta\overline{v}$ is called the scalar multiplication of the interval-valued cooperative game $\overline{v} \in \overline{G}^n$ and the positive real number β. Clearly, $\beta\overline{v}$ is still an interval-valued cooperative game belonging to $\overline{G}^n$, i.e., $\beta\overline{v} \in \overline{G}^n$.

Analogously, for any interval-valued cooperative games $\overline{v} \in \overline{G}^n$ and $\overline{\nu} \in \overline{G}^n$, according to the case 2 of Definition 1.1, $\overline{v} + \overline{\nu}$ is defined as an interval-valued cooperative game with the interval-valued characteristic function $\overline{v} + \overline{\nu}$, where

$$(\overline{v} + \overline{\nu})(S) = \overline{v}(S) + \overline{\nu}(S)$$

for any coalition $S \subseteq N$, i.e.,

$$(\overline{v} + \overline{\nu})(S) = [v_L(S) + \nu_L(S), v_R(S) + \nu_R(S)]. \tag{1.7}$$

Usually, $\overline{v} + \overline{\nu}$ is called the sum of the interval-valued cooperative games $\overline{v} \in \overline{G}^n$ and $\overline{\nu} \in \overline{G}^n$. Obviously, $\overline{v} + \overline{\nu}$ is also an interval-valued cooperative game belonging to $\overline{G}^n$, i.e., $(\overline{v} + \overline{\nu}) \in \overline{G}^n$.

For any interval-valued cooperative game $\overline{v} \in \overline{G}^n$, it is easy to see that each player should receive an interval-valued payoff from the cooperation due to the fact that each coalition's value is an interval. Let $\overline{x}_i(\overline{v}) = [x_{Li}(\overline{v}), x_{Ri}(\overline{v})]$ be the interval-valued payoff which is allocated to the player $i \in N$ under the cooperation that the grand coalition is reached. Denote $\overline{\boldsymbol{x}}(\overline{v}) = (\overline{x}_1(\overline{v}), \overline{x}_2(\overline{v}), \ldots, \overline{x}_n(\overline{v}))^{\mathrm{T}}$, which is the vector of the interval-valued payoffs for all n players in the grand coalition N. For any coalition $S \subseteq N$, denote

$$\overline{x}(\overline{v}, S) = \sum_{i \in S} \overline{x}_i(\overline{v}),$$

which represents the sum of the interval-valued payoffs of all the players in the coalition S. According to the case 2 of Definition 1.1 (i.e., the interval addition [15]), we can express $\overline{x}(\overline{v}, S)$ as the following interval:

$$\overline{x}(\overline{v}, S) = \left[\sum_{i \in S} x_{Li}(\overline{v}), \sum_{i \in S} x_{Ri}(\overline{v}) \right].$$

In a similar way to the definitions of the efficiency and individual rationality stated as in Sect. 1.2, for an interval-valued cooperative game $\overline{v} \in \overline{G}^n$, the efficiency and individual rationality of an interval-valued payoff vector $\overline{\boldsymbol{x}}(\overline{v}) = (\overline{x}_1(\overline{v}), \overline{x}_2(\overline{v}), \ldots, \overline{x}_n(\overline{v}))^{\mathrm{T}}$ can be expressed as follows:

$$\sum_{i=1}^{n} \overline{x}_i(\overline{v}) = \overline{v}(N)$$

and

$$\overline{x}_i(\overline{v}) \geq \overline{v}(i) \quad (i = 1, 2, \ldots, n),$$

respectively. It is easily derived from Eq. (1.4) (i.e., the Moore's order relation over intervals [15]) that

$$\begin{cases} \sum_{i=1}^{n} x_{Li}(\overline{v}) = v_L(N) \\ \sum_{i=1}^{n} x_{Ri}(\overline{v}) = v_R(N) \end{cases} \tag{1.8}$$

and

$$\begin{cases} x_{Li}(\overline{v}) \geq v_L(i) \;\; (i = 1, 2, \ldots, n) \\ x_{Ri}(\overline{v}) \geq v_R(i) \;\; (i = 1, 2, \ldots, n). \end{cases} \tag{1.9}$$

A vector $\overline{\boldsymbol{x}}(\overline{v})$ of the interval-valued payoffs is called an interval-valued imputation of the interval-valued cooperative game $\overline{v} \in \overline{G}^n$ if it satisfies the efficiency and individual rationality. Denote the set of interval-valued imputations of an interval-valued cooperative game $\overline{v} \in \overline{G}^n$ by $\overline{I}(\overline{v})$. Generally, an interval-valued imputation is very large. Thus, we will further study the interval-valued core of interval-valued cooperative games in the subsequent Chap. 2.

To measure the difference between the intervals $\overline{x}(\overline{v}, S)$ and $\overline{v}(S)$, we use the concept of the distance between intervals. Thus, according to Eq. (1.5), we define the square of the distance between the intervals $\overline{x}(\overline{v}, S)$ and $\overline{v}(S)$ for the coalition S as follows:

$$D(\overline{x}(\overline{v}, S), \overline{v}(S)) = \left(\sum_{i \in S} x_{Li}(\overline{v}) - v_L(S)\right)^2 + \left(\sum_{i \in S} x_{Ri}(\overline{v}) - v_R(S)\right)^2.$$

Accordingly, the sum of the squares of the distances between the intervals $\overline{x}(\overline{v}, S)$ and $\overline{v}(S)$ for all coalitions $S \subseteq N$ can be defined as follows:

$$L(\overline{\boldsymbol{x}}(\overline{v})) = \sum_{S \subseteq N} D\left(\overline{x}(\overline{v}, S), \overline{v}(S)\right),$$

which directly implies that

$$L(\overline{\boldsymbol{x}}(\overline{v})) = \sum_{S \subseteq N} \left[\left(\sum_{i \in S} x_{Li}(\overline{v}) - v_L(S)\right)^2 + \left(\sum_{i \in S} x_{Ri}(\overline{v}) - v_R(S)\right)^2\right].$$

From a view of players' profit allocation, $L(\overline{\boldsymbol{x}}(\overline{v}))$ may be interpreted as a loss function. Thus, a solution of an interval valued cooperative game $\overline{v} \in \overline{G}^n$ can be defined as the interval-valued payoff vector $\overline{x}^*(\overline{v}) = (\overline{x}_1^*(\overline{v}), \overline{x}_2^*(\overline{v}), \ldots, \overline{x}_n^*(\overline{v}))^{\mathrm{T}}$, which is the optimal solution of the quadratic programming model as follows:

$$\min\{L(\overline{x}(\overline{v}))\},$$

i.e.,

$$\min\left\{\sum_{S\subseteq N}\left[\left(\sum_{i\in S}x_{Li}(\overline{v})-v_L(S)\right)^2+\left(\sum_{i\in S}x_{Ri}(\overline{v})-v_R(S)\right)^2\right]\right\}. \tag{1.10}$$

Such a solution $\overline{x}^*(\overline{v})$ is called the interval-valued least square solution of the interval-valued cooperative game $\overline{v} \in \overline{G}^n$, denoted by $\overline{\rho}^{\mathrm{LS}}(\overline{v})$, i.e., $\overline{\rho}^{\mathrm{LS}}(\overline{v}) = \overline{x}^*(\overline{v})$. The term "least square" is originated from the least square method which is applied to Eq. (1.10). In other words, the interval-valued least square solution of the interval-valued cooperative game $\overline{v} \in \overline{G}^n$ is to minimize the loss function $L(\overline{x}(\overline{v}))$ in the sense of the least square distance.

1.4 Quadratic Programming Models and Methods for the Interval-Valued Least Square Solutions of Interval-Valued Cooperative Games

In this section, we focus on developing two effective and fast quadratic programming methods for solving interval-valued cooperative games as stated in Sect. 1.3.2.

1.4.1 *The Interval-Valued Least Square Solution Without Considering the Efficiency*

It is easy to see from Eq. (1.10) that computing the interval-valued least square solution of an interval-valued cooperative game $\overline{v} \in \overline{G}^n$ becomes solving the quadratic programming model in the sense of minimizing the coalitions' loss. If the efficiency is not considered for the moment, then Eq. (1.10) is in fact an unconstrained optimization problem.

Partial derivatives of $L(\overline{x}(\overline{v}))$ with respect to the variables $x_{Lj}(\overline{v})$ and $x_{Rj}(\overline{v})$ $(j \in S \subseteq N)$ are computed as follows:

$$\frac{\partial L(\overline{x}(\overline{v}))}{\partial x_{Lj}(\overline{v})} = 2\sum_{S\subseteq N:\, j\in S}\left(\sum_{i\in S} x_{Li}(\overline{v}) - v_L(S)\right) \quad (j = 1, 2, \ldots, n)$$

and

$$\frac{\partial L(\overline{x}(\overline{v}))}{\partial x_{Rj}(\overline{v})} = 2\sum_{S\subseteq N:\, j\in S}\left(\sum_{i\in S} x_{Ri}(\overline{v}) - v_R(S)\right) \quad (j = 1, 2, \ldots, n),$$

respectively.

Let the partial derivatives of $L(\overline{x}(\overline{v}))$ with respect to the variables $x_{Lj}(\overline{v})$ and $x_{Rj}(\overline{v})$ $(j \in S \subseteq N)$ be equal to 0, respectively. Thus, we have

$$2\sum_{S\subseteq N:\, j\in S}\left(\sum_{i\in S} x^*_{Li}(\overline{v}) - v_L(S)\right) = 0 \quad (j = 1, 2, \ldots, n)$$

and

$$2\sum_{S\subseteq N:\, j\in S}\left(\sum_{i\in S} x^*_{Ri}(\overline{v}) - v_R(S)\right) \quad (j = 1, 2, \ldots, n),$$

respectively, which directly infer that

$$\sum_{S\subseteq N:\, j\in S}\sum_{i\in S} x^*_{Li}(\overline{v}) = \sum_{S\subseteq N: j\in S} v_L(S) \quad (j = 1, 2, \ldots, n) \tag{1.11}$$

and

$$\sum_{S\subseteq N:\, j\in S}\sum_{i\in S} x^*_{Ri}(\overline{v}) = \sum_{S\subseteq N:\, j\in S} v_R(S) \quad (j = 1, 2, \ldots, n). \tag{1.12}$$

Solving the above systems of linear equations (i.e., Eqs. (1.11) and (1.12)), we can obtain the interval-valued least square solution of the interval-valued cooperative game $\overline{v} \in \overline{G}^n$. Thus, in what follows, we focus on how to solve Eqs. (1.11) and (1.12).

To solve $x^*_{Li}(\overline{v})$ $(i = 1, 2, \ldots, n)$ and $x^*_{Ri}(\overline{v})$ $(i = 1, 2, \ldots, n)$, Eqs. (1.11) and (1.12) can be rewritten as follows:

$$\begin{cases} a_{11}x^*_{L1}(\overline{v}) + a_{12}x^*_{L2}(\overline{v}) + \cdots + a_{1n}x^*_{Ln}(\overline{v}) = \sum\limits_{S\subseteq N: 1\in S} v_L(S) \\ a_{21}x^*_{L1}(\overline{v}) + a_{22}x^*_{L2}(\overline{v}) + \cdots + a_{2n}x^*_{Ln}(\overline{v}) = \sum\limits_{S\subseteq N: 2\in S} v_L(S) \\ \cdots \\ a_{n1}x^*_{L1}(\overline{v}) + a_{n2}x^*_{L2}(\overline{v}) + \cdots + a_{nn}x^*_{Ln}(\overline{v}) = \sum\limits_{S\subseteq N: n\in S} v_L(S) \end{cases} \tag{1.13}$$

and

$$\begin{cases} a_{11}x_{R1}^{*}(\overline{v}) + a_{12}x_{R2}^{*}(\overline{v}) + \cdots + a_{1n}x_{Rn}^{*}(\overline{v}) = \sum\limits_{S\subseteq N:1\in S} v_R(S) \\ a_{21}x_{R1}^{*}(\overline{v}) + a_{22}x_{R2}^{*}(\overline{v}) + \cdots + a_{2n}x_{Rn}^{*}(\overline{v}) = \sum\limits_{S\subseteq N:2\in S} v_R(S) \\ \cdots \\ a_{n1}x_{R1}^{*}(\overline{v}) + a_{n2}x_{R2}^{*}(\overline{v}) + \cdots + a_{nn}x_{Rn}^{*}(\overline{v}) = \sum\limits_{S\subseteq N:n\in S} v_R(S), \end{cases} \tag{1.14}$$

respectively.

Let s be the number of all players in the coalition $S \subseteq N$. As stated earlier, s is in fact the cardinality of the coalition $S \subseteq N$, i.e., $s = |S|$. According to the knowledge on the theory of permutation and combination, for the player $i \in N$, the number of the coalitions S including the player i with $s = 1$ can be expressed as C_{n-1}^{0}. In the same way, the number of the coalitions S including the player i with $s = 2$ can be expressed as C_{n-1}^{1}. Generally, the number of the coalitions S including the player i with $s = k(k = 1, 2, \ldots, n)$ can be expressed as C_{n-1}^{k-1}. It is obvious that the number of the coalitions S including the player i can be written as

$$C_{n-1}^{0} + C_{n-1}^{1} + \cdots + C_{n-1}^{n-2} + C_{n-1}^{n-1},$$

which is equal to 2^{n-1} by the simple observation, where C_{n-1}^{k-1} $(k = 1, 2, \ldots, n)$ is the combination which is computed as follows:

$$C_{n-1}^{k-1} = \frac{(n-1)!}{(k-1)!(n-k)!}.$$

Similarly, for the players $i \in N$ and $j \in N$ $(i \neq j)$, the number of the coalitions S including both players i and j with $s = 2$ can be expressed as C_{n-2}^{0}, the number of the coalitions S including both players i and j with $s = 3$ can be expressed as C_{n-2}^{1}. Generally, the number of the coalitions S including both players i and j with $s = k$ $(k = 2, 3, \ldots, n)$ can be expressed as C_{n-2}^{k-2}. Thus, the number of the coalitions S including both players i and j can be written as

$$C_{n-2}^{0} + C_{n-2}^{1} + \cdots + C_{n-2}^{n-3} + C_{n-2}^{n-2},$$

which is equal to 2^{n-2}.

Thus, it follows from the aforementioned conclusions that

$$a_{ij} = \begin{cases} 2^{n-1} & \text{if } i \in N, j \in N \text{ and } i = j \\ 2^{n-2} & \text{if } i \in N, j \in N \text{ and } i \neq j. \end{cases}$$

Denote $\boldsymbol{x}_L^*(\overline{v}) = \left(x_{L1}^*(\overline{v}), x_{L2}^*(\overline{v}), \ldots, x_{Ln}^*(\overline{v})\right)^{\mathrm{T}}, \boldsymbol{x}_R^*(\overline{v}) = \left(x_{R1}^*(\overline{v}),\ x_{R2}^*(\overline{v}), \ldots, x_{Rn}^*(\overline{v})\right)^{\mathrm{T}}$,

$$\boldsymbol{b}_L(\overline{v}) = \left(\sum_{S \subseteq N:1 \in S} v_L(S), \sum_{S \subseteq N:2 \in S} v_L(S), \ldots, \sum_{S \subseteq N:n \in S} v_L(S)\right)^{\mathrm{T}}, \tag{1.15}$$

$$\boldsymbol{b}_R(\overline{v}) = \left(\sum_{S \subseteq N:1 \in S} v_R(S), \sum_{S \subseteq N:2 \in S} v_R(S), \ldots, \sum_{S \subseteq N:n \in S} v_R(S)\right)^{\mathrm{T}}, \tag{1.16}$$

and

$$\boldsymbol{A} = \left(a_{ij}\right)_{n \times n} = \begin{pmatrix} 2^{n-1} & 2^{n-2} & \cdots & 2^{n-2} \\ 2^{n-2} & 2^{n-1} & \cdots & 2^{n-2} \\ \vdots & \vdots & & \vdots \\ 2^{n-2} & 2^{n-2} & \cdots & 2^{n-1} \end{pmatrix}_{n \times n}. \tag{1.17}$$

Accordingly, Eqs. (1.13) and (1.14) can be rewritten in the matrix format as follows:

$$\boldsymbol{A}\boldsymbol{x}_L^*(\overline{v}) = \boldsymbol{b}_L(\overline{v}) \tag{1.18}$$

and

$$\boldsymbol{A}\boldsymbol{x}_R^*(\overline{v}) = \boldsymbol{b}_R(\overline{v}), \tag{1.19}$$

respectively.

Let

$$(\boldsymbol{A}, \boldsymbol{E}) = \begin{pmatrix} 2^{n-1} & 2^{n-2} & \cdots & 2^{n-2} & 1 & 0 & \cdots & 0 \\ 2^{n-2} & 2^{n-1} & \cdots & 2^{n-2} & 0 & 1 & \cdots & 0 \\ \vdots & \vdots & & \vdots & \vdots & \vdots & & \vdots \\ 2^{n-2} & 2^{n-2} & \cdots & 2^{n-1} & 0 & 0 & & 1 \end{pmatrix}_{n \times 2n},$$

where $\boldsymbol{E}$ is the identity matrix, i.e.,

$$\boldsymbol{E} = \begin{pmatrix} 1 & 0 & \cdots & 0 \\ 0 & 1 & \cdots & 0 \\ \vdots & \vdots & & \vdots \\ 0 & 0 & \cdots & 1 \end{pmatrix}_{n \times n}.$$

By using the elementary linear transformation, we have

$$(\boldsymbol{A},\boldsymbol{E}) \sim \begin{pmatrix} 1 & 0 & \cdots & 0 & \frac{1}{2^{n-2}}\times\frac{n}{n+1} & -\frac{1}{2^{n-2}}\times\frac{1}{n+1} & \cdots & -\frac{1}{2^{n-2}}\times\frac{1}{n+1} \\ 0 & 1 & \cdots & 0 & -\frac{1}{2^{n-2}}\times\frac{1}{n+1} & \frac{1}{2^{n-2}}\times\frac{n}{n+1} & \cdots & -\frac{1}{2^{n-2}}\times\frac{1}{n+1} \\ \vdots & \vdots & & \vdots & \vdots & \vdots & & \vdots \\ 0 & 0 & & 1 & -\frac{1}{2^{n-2}}\times\frac{1}{n+1} & -\frac{1}{2^{n-2}}\times\frac{1}{n+1} & \cdots & \frac{1}{2^{n-2}}\times\frac{n}{n+1} \end{pmatrix}_{n\times 2n}.$$

Clearly, the matrixes $\boldsymbol{A}$ and $\boldsymbol{E}$ are row equivalent. Therefore, the matrix $\boldsymbol{A}$ is reversible. Hereby, we have

$$\boldsymbol{A}^{-1} = \begin{pmatrix} \frac{1}{2^{n-2}}\times\frac{n}{n+1} & -\frac{1}{2^{n-2}}\times\frac{1}{n+1} & \cdots & -\frac{1}{2^{n-2}}\times\frac{1}{n+1} \\ -\frac{1}{2^{n-2}}\times\frac{1}{n+1} & \frac{1}{2^{n-2}}\times\frac{n}{n+1} & \cdots & -\frac{1}{2^{n-2}}\times\frac{1}{n+1} \\ \vdots & \vdots & & \vdots \\ -\frac{1}{2^{n-2}}\times\frac{1}{n+1} & -\frac{1}{2^{n-2}}\times\frac{1}{n+1} & \cdots & \frac{1}{2^{n-2}}\times\frac{n}{n+1} \end{pmatrix}_{n\times n},$$

i.e.,

$$\boldsymbol{A}^{-1} = \frac{1}{2^{n-2}} \begin{pmatrix} \frac{n}{n+1} & -\frac{1}{n+1} & \cdots & -\frac{1}{n+1} \\ -\frac{1}{n+1} & \frac{n}{n+1} & \cdots & -\frac{1}{n+1} \\ \vdots & \vdots & & \vdots \\ -\frac{1}{n+1} & -\frac{1}{n+1} & \cdots & \frac{n}{n+1} \end{pmatrix}_{n\times n}. \tag{1.20}$$

By using the multiplication of matrixes, we obtain the solutions of Eqs. (1.18) and (1.19) (i.e., Eqs. (1.13) and (1.14)) as follows:

$$\boldsymbol{x}_L^*(\overline{\upsilon}) = \boldsymbol{A}^{-1}\boldsymbol{b}_L(\overline{\upsilon}) \tag{1.21}$$

and

$$\boldsymbol{x}_R^*(\overline{\upsilon}) = \boldsymbol{A}^{-1}\boldsymbol{b}_R(\overline{\upsilon}), \tag{1.22}$$

respectively. Thus, we obtain the interval-valued least square solution $\overline{\boldsymbol{\rho}}^{\mathrm{LS}}(\overline{\upsilon})$ of the interval-valued cooperative game $\overline{\upsilon} \in \overline{G}^n$, i.e., $\overline{\boldsymbol{\rho}}^{\mathrm{LS}}(\overline{\upsilon}) = \overline{\boldsymbol{x}}^*(\overline{\upsilon})$, whose components are expressed as the intervals $\overline{x}_i^*(\overline{\upsilon}) = \left[x_{Li}^*(\overline{\upsilon}), x_{Ri}^*(\overline{\upsilon})\right]$ $(i = 1, 2, \ldots, n)$, which are specified as follows.

According to Eqs. (1.21) and (1.22), we obtain

$$x^*_{Li}(\overline{v}) = \boldsymbol{A}^{-1}_{i.}\boldsymbol{b}_L(\overline{v})$$

and

$$x^*_{Ri}(\overline{v}) = \boldsymbol{A}^{-1}_{i.}\boldsymbol{b}_R(\overline{v}),$$

where $\boldsymbol{A}^{-1}_{i.}$ is the ith row of the matrix $\boldsymbol{A}^{-1}$ given by Eq. (1.20), i.e.,

$$\boldsymbol{A}^{-1}_{i.} = \frac{1}{2^{n-2}}\left(\overbrace{-\frac{1}{n+1}\ -\frac{1}{n+1}\ \cdots\ -\frac{1}{n+1}}^{i-1}\ \overbrace{\frac{n}{n+1}}^{i\text{th}}\ \overbrace{-\frac{1}{n+1}\ -\frac{1}{n+1}\ \cdots\ -\frac{1}{n+1}}^{n-i}\right). \tag{1.23}$$

Combining with Eqs. (1.15) and (1.16), we obtain

$$x^*_{Li}(\overline{v}) = \frac{1}{2^{n-2}(n+1)} \times \Big(-\sum_{S\subseteq N:1\in S} v_L(S) - \cdots - \sum_{S\subseteq N:i-1\in S} v_L(S) + n\sum_{S\subseteq N:i\in S} v_L(S) - \sum_{S\subseteq N:i+1\in S} v_L(S) - \cdots - \sum_{S\subseteq N:n\in S} v_L(S)\Big)$$

and

$$x^*_{Ri}(\overline{v}) = \frac{1}{2^{n-2}(n+1)} \times \Big(-\sum_{S\subseteq N:1\in S} v_R(S) - \cdots - \sum_{S\subseteq N:i-1\in S} v_R(S) + n\sum_{S\subseteq N:i\in S} v_R(S) - \sum_{S\subseteq N:i+1\in S} v_R(S) - \cdots - \sum_{S\subseteq N:n\in S} v_R(S)\Big)$$

which can be rewritten as follows:

$$x^*_{Li}(\overline{v}) = \frac{n\sum_{S\subseteq N:i\in S} v_L(S) - \sum_{j=1,j\neq i}^{n}\sum_{S\subseteq N:j\in S} v_L(S)}{2^{n-2}(n+1)} \tag{1.24}$$

and

$$x^*_{Ri}(\overline{v}) = \frac{n\sum_{S\subseteq N:i\in S} v_R(S) - \sum_{j=1,j\neq i}^{n}\sum_{S\subseteq N:j\in S} v_R(S)}{2^{n-2}(n+1)}, \tag{1.25}$$

respectively.

In what follows, we discuss some useful and important properties of the interval-valued least square solutions of interval-valued cooperative games.

Theorem 1.2 *(Existence and Uniqueness) For an arbitrary interval-valued cooperative game $\overline{\upsilon} \in \overline{G}^n$, there always exists a unique interval-valued least square solution $\overline{\rho}^{\mathrm{LS}}(\overline{\upsilon})$, which is determined by Eqs. (1.21) and (1.22) (or Eqs. (1.24) and (1.25)).*

Proof According to Eqs. (1.21) and (1.22) (or Eqs. (1.24) and (1.25)), it is straightforward to prove Theorem 1.2.

Theorem 1.3 *(Additivity) For any two interval-valued cooperative games $\overline{\upsilon} \in \overline{G}^n$ and $\overline{\nu} \in \overline{G}^n$, then $\overline{x}_i^*(\overline{\upsilon}+\overline{\nu}) = \overline{x}_i^*(\overline{\upsilon}) + \overline{x}_i^*(\overline{\nu})$ $(i = 1, 2, \ldots, n)$, i.e., $\overline{\rho}^{\mathrm{LS}}(\overline{\upsilon}+\overline{\nu}) = \overline{\rho}^{\mathrm{LS}}(\overline{\upsilon}) + \overline{\rho}^{\mathrm{LS}}(\overline{\nu})$.*

Proof According to Eq. (1.24), we have

$$\begin{aligned}
x_{Li}^*(\overline{\upsilon}+\overline{\nu}) &= \frac{n\sum\limits_{S\subseteq N:i\in S}(\upsilon_L(S)+\nu_L(S)) - \sum\limits_{j=1,j\neq i}^{n}\sum\limits_{S\subseteq N:j\in S}(\upsilon_L(S)+\nu_L(S))}{2^{n-2}(n+1)} \\
&= \frac{n\sum\limits_{S\subseteq N:i\in S}\upsilon_L(S) - \sum\limits_{j=1,j\neq i}^{n}\sum\limits_{S\subseteq N:j\in S}\upsilon_L(S)}{2^{n-2}(n+1)} \\
&\quad + \frac{n\sum\limits_{S\subseteq N:i\in S}\nu_L(S) - \sum\limits_{j=1,j\neq i}^{n}\sum\limits_{S\subseteq N:j\in S}\nu_L(S)}{2^{n-2}(n+1)} \\
&= x_{Li}^*(\overline{\upsilon}) + x_{Li}^*(\overline{\nu}),
\end{aligned}$$

i.e.,

$$x_{Li}^*(\overline{\upsilon}+\overline{\nu}) = x_{Li}^*(\overline{\upsilon}) + x_{Li}^*(\overline{\nu}).$$

Analogously, according to Eq. (1.25), we can easily prove that

$$x_{Ri}^*(\overline{\upsilon}+\overline{\nu}) = x_{Ri}^*(\overline{\upsilon}) + x_{Ri}^*(\overline{\nu}).$$

Combining with the aforementioned conclusion, according to the case 1 of Definition 1.1, we obtain

$$\overline{x}_i^*(\overline{\upsilon}+\overline{\nu}) = \overline{x}_i^*(\overline{\upsilon}) + \overline{x}_i^*(\overline{\nu}) \quad (i = 1, 2, \ldots, n).$$

Namely,

$$\overline{\rho}^{\mathrm{LS}}(\overline{\upsilon}+\overline{\nu}) = \overline{\rho}^{\mathrm{LS}}(\overline{\upsilon}) + \overline{\rho}^{\mathrm{LS}}(\overline{\nu}).$$

Thus, we have proven Theorem 1.3

Definition 1.3 (Symmetric player) For two players $i \in N$ and $k \in N(i \neq k)$, if

$$\overline{\upsilon}(S \cup i) = \overline{\upsilon}(S \cup k)$$

for any coalition $S \subseteq N \backslash \{i, k\}$, then the players i and k are said to be symmetric in the interval-valued cooperative game $\overline{\upsilon} \in \overline{G}^n$.

Clearly, two symmetric players have the identical contribution to any coalition. Consequently, it seems to be reasonable that two symmetric players in the interval-valued cooperative game should obtain the identical payoff according to the interval-valued least square solution. Formally, this is what the following Theorem 1.4 states.

Theorem 1.4 *(Symmetry) If $i \in N$ and $k \in N(i \neq k)$ are two symmetric players in an interval-valued cooperative game $\overline{\upsilon} \in \overline{G}^n$, then $\overline{x}_i^*(\overline{\upsilon}) = \overline{x}_k^*(\overline{\upsilon})$, i.e., $\overline{\rho}_i^{\mathrm{LS}}(\overline{\upsilon}) = \overline{\rho}_k^{\mathrm{LS}}(\overline{\upsilon})$.*

Proof For the players $i \in N$ and $k \in N(i \neq k)$, according to Eq. (1.24), we have

$$x_{Li}^*(\overline{\upsilon}) = \frac{-\sum_{j=1, j \neq i, j \neq k}^{n} \sum_{S \subseteq N: j \in S} \upsilon_L(S) + \left(n \sum_{S \subseteq N: i \in S} \upsilon_L(S) - \sum_{S \subseteq N: k \in S} \upsilon_L(S) \right)}{2^{n-2}(n+1)} \tag{1.26}$$

and

$$x_{Lk}^*(\overline{\upsilon}) = \frac{-\sum_{j=1, j \neq i, j \neq k}^{n} \sum_{S \subseteq N: j \in S} \upsilon_L(S) + \left(-\sum_{S \subseteq N: i \in S} \upsilon_L(S) + n \sum_{S \subseteq N: k \in S} \upsilon_L(S) \right)}{2^{n-2}(n+1)}. \tag{1.27}$$

Due to the assumption that the players i and k are symmetric in the interval-valued cooperative game $\overline{\upsilon} \in \overline{G}^n$, it easily follows from Definition 1.3 and Eqs. (1.26) and (1.27) that

$$\sum_{S \subseteq N: i \in S} \upsilon_L(S) = \sum_{S \subseteq N: k \in S} \upsilon_L(S),$$

which directly infers that

$$n \sum_{S \subseteq N: i \in S} \upsilon_L(S) - \sum_{S \subseteq N: k \in S} \upsilon_L(S) = -\sum_{S \subseteq N: i \in S} \upsilon_L(S) + n \sum_{S \subseteq N: k \in S} \upsilon_L(S).$$

Note that $\sum_{j=1, j \neq i, j \neq k}^{n} \sum_{S \subseteq N: j \in S} \upsilon_L(S)$ in Eqs. (1.26) and (1.27) is independent of the players i and k. Hereby, it follows from Eqs. (1.26) and (1.27) that $x_{Li}^*(\overline{\upsilon}) = x_{Lk}^*(\overline{\upsilon})$.

In the same way, according to Eq. (1.25), we can prove $x_{Ri}^*(\overline{v}) = x_{Rk}^*(\overline{v})$. Combining with the aforementioned conclusion and the case 1 of Definition 1.1, we can obtain

$$[x_{Li}^*(\overline{v}), x_{Ri}^*(\overline{v})] = [x_{Lk}^*(\overline{v}), x_{Rk}^*(\overline{v})],$$

i.e.,

$$\overline{x}_i^*(\overline{v}) = \overline{x}_k^*(\overline{v})$$

or

$$\overline{\rho}_i^{\mathrm{LS}}(\overline{v}) = \overline{\rho}_k^{\mathrm{LS}}(\overline{v}).$$

Accordingly, we have completed the proof of Theorem 1.4.

Definition 1.4 (Null player) For a player $i \in N$, if

$$\overline{v}(S \cup i) = \overline{v}(S)$$

for any coalition $S \subseteq N \backslash i$, then i is called a null player in the interval-valued cooperative game $\overline{v} \in \overline{G}^n$.

A null player does not contribute anything to any coalition, particularly $\overline{v}(i) = 0$. Thus, it seems to be reasonable that a null player in the interval-valued cooperative game obtains zero according to the interval-valued least square solution. Formally, this is what the following Theorem 1.5 states.

Theorem 1.5 *(Null player) If $i \in N$ is a null player in an interval-valued cooperative game $\overline{v} \in \overline{G}^n$, then $\overline{x}_i^*(\overline{v}) = 0$, i.e., $\overline{\rho}_i^{\mathrm{LS}}(\overline{v}) = 0$.*

Proof According to Eq. (1.24), we have

$$\begin{aligned} x_{Li}^*(\overline{v}) &= \frac{n \sum\limits_{S \subseteq N: i \in S} v_L((S \backslash i) \cup i) - \sum\limits_{j=1, j \neq i}^{n} \sum\limits_{S \subseteq N: j \in S} v_L(S)}{2^{n-2}(n+1)} \\ &= \frac{n \sum\limits_{S \subseteq N: i \in S} v_L(S \backslash i) - \sum\limits_{j=1, j \neq i}^{n} \sum\limits_{S \subseteq N: j \in S} v_L(S)}{2^{n-2}(n+1)} \end{aligned}$$

due to the assumption that i is a null player in the interval-valued cooperative game $\overline{v} \in \overline{G}^n$. Hereby, we have $x_{Li}^*(\overline{v}) = 0$.

Analogously, according to Eq. (1.25), we can prove $x_{Ri}^*(\overline{v}) = 0$. Thereby, we obtain

$$[x_{Li}^*(\overline{v}), x_{Ri}^*(\overline{v})] = 0,$$

i.e., $\overline{x}_i^*(\overline{v}) = 0$ or $\overline{\rho}_i^{\mathrm{LS}}(\overline{v}) = 0$. Thus, we have proven Theorem 1.5.

Definition 1.5 (Dummy player) For a player $i \in N$, if

$$\overline{v}(S \cup i) = \overline{v}(S) + \overline{v}(i)$$

for any coalition $S \subseteq N \backslash i$, then i is called a dummy player in the interval-valued cooperative game $\overline{v} \in \overline{G}^n$.

Obviously, a dummy player i only contributes his/her own worth $\overline{v}(i)$ to every coalition. Thus, it seems to be reasonable that a dummy player i in the interval-valued cooperative game should obtain his/her own worth $\overline{v}(i)$ according to the interval-valued least square solution. Formally, this is what the following Theorem 1.6 states.

Theorem 1.6 *(Dummy player) If $i \in N$ is a dummy player in an interval-valued cooperative game $\overline{v} \in \overline{G}^n$, then $\overline{x}_i^*(\overline{v}) = \overline{v}(i)$, i.e., $\overline{\rho}_i^{\mathrm{LS}}(\overline{v}) = \overline{v}(i)$.*

Proof It can be easily proven in a very similar way to that of Theorem 1.5 (omitted).

Let σ be any permutation on the set N. For an interval-valued cooperative game $\overline{v} \in \overline{G}^n$, we can define the interval-valued cooperative game $\overline{v}^\sigma \in \overline{G}^n$ with interval-valued characteristic function $\overline{v}^\sigma$, where $\overline{v}^\sigma(S) = \overline{v}(\sigma^{-1}(S))$ for any coalition $S \subseteq N$.

Let $\boldsymbol{\sigma}^\# : R^n \mapsto R^n$ be a mapping so that for any vector $\boldsymbol{z} = (z_1, z_2, \ldots, z_n)^{\mathrm{T}} \in R^n$ and $i \in N$,

$$\sigma_{\sigma(i)}^\#(\boldsymbol{z}) = z_i,$$

where $\boldsymbol{\sigma}^\#(\boldsymbol{z}) = \left(\sigma_{\sigma(1)}^\#(\boldsymbol{z}), \sigma_{\sigma(2)}^\#(\boldsymbol{z}), \ldots, \sigma_{\sigma(n)}^\#(\boldsymbol{z})\right)^{\mathrm{T}}$.

Theorem 1.7 *(Anonymity) For any permutation σ on the set N and an interval-valued cooperative game $\overline{v} \in \overline{G}^n$, then $\overline{x}_{\sigma(i)}^*(\overline{v}^\sigma) = \overline{x}_i^*(\overline{v})$, i.e., $\overline{\rho}_{\sigma(i)}^{\mathrm{LS}}(\overline{v}^\sigma) = \overline{\rho}_i^{\mathrm{LS}}(\overline{v})$. Namely, $\overline{\boldsymbol{\rho}}^{\mathrm{LS}}(\overline{v}^\sigma) = \boldsymbol{\sigma}^\#\left(\overline{\boldsymbol{\rho}}^{\mathrm{LS}}(\overline{v})\right)$.*

Proof It can be easily proven according to Eqs. (1.24) and (1.25) (omitted).

It is easy to see from Theorem 1.7 that the interval-valued least square solution of any interval-valued cooperative game satisfies the anonymity, i.e., $\overline{\boldsymbol{\rho}}^{\mathrm{LS}}(\overline{v}^\sigma) = \boldsymbol{\sigma}^\#\left(\overline{\boldsymbol{\rho}}^{\mathrm{LS}}(\overline{v})\right)$. The anonymity implies that the interval-valued least square solution does not discriminate between the players solely on the basis of their "names," i.e., numbers.

From the above discussion, it is obvious that the dummy player property implies the null player property, and the anonymity implies the symmetry. In other words, the dummy player property and the anonymity are stronger versions of the null player property and the symmetry, respectively.

In addition, if all coalitions' values $\overline{v}(S)$ degenerate to real numbers, i.e., $v(S) = v_L(S) = v_R(S)$ for any coalition $S \subseteq N$, then it easily follows from Eqs. (1.15) and (1.16) that $\boldsymbol{b}(v) = \boldsymbol{b}_L(\overline{v}) = \boldsymbol{b}_R(\overline{v})$. Hereby, Eqs. (1.21) and (1.22) are identical.

That is to say, Eqs. (1.21) and (1.22) are applicable to the classical cooperative games. Thus, the models and method developed in this subsection may be regarded as an extension of that for the classical cooperative games when uncertainty and imprecision are taken into consideration.

1.4.2 The Interval-Valued Least Square Solution with Considering the Efficiency

In real management situations, some constraint conditions have to be taken into consideration. In this case, the quadratic programming model (i.e., Eq. (1.10)) is still applicable. For example, if we consider the efficiency condition:

$$\overline{x}(\overline{v},N)=\overline{v}(N),$$

i.e.,

$$\left[\sum_{i=1}^{n}x_{Li}(\overline{v}),\sum_{i=1}^{n}x_{Ri}(\overline{v})\right]=[v_L(N),v_R(N)],$$

then Eq. (1.10) can be flexibly rewritten as the following quadratic programming model:

$$\min\left\{\sum_{S\subseteq N}\left[\left(\sum_{i\in S}x_{Li}(\overline{v})-v_L(S)\right)^2+\left(\sum_{i\in S}x_{Ri}(\overline{v})-v_R(S)\right)^2\right]\right\}$$
$$\text{s.t.}\begin{cases}\sum_{i=1}^{n}x_{Li}(\overline{v})=v_L(N)\\ \sum_{i=1}^{n}x_{Ri}(\overline{v})=v_R(N).\end{cases}\tag{1.28}$$

Stated as earlier, the optimal solution $\overline{\boldsymbol{x}}^{*\mathrm{E}}(\overline{v})=\left(\overline{x}_1^{*\mathrm{E}}(\overline{v}),\overline{x}_2^{*\mathrm{E}}(\overline{v}),\ldots,\overline{x}_n^{*\mathrm{E}}(\overline{v})\right)^{\mathrm{T}}$ of Eq. (1.28) is called the interval-valued least square solution with considering the efficiency for the interval-valued cooperative game $\overline{v}\in\overline{G}^n$, denoted by $\overline{\boldsymbol{\rho}}^{\mathrm{LSE}}(\overline{v})$, i.e., $\overline{\boldsymbol{\rho}}^{\mathrm{LSE}}(\overline{v})=\overline{\boldsymbol{x}}^{*\mathrm{E}}(\overline{v})$, where $\overline{x}_i^{*\mathrm{E}}(\overline{v})=\left[x_i^{*\mathrm{E}}(\overline{v}),x_i^{*\mathrm{E}}(\overline{v})\right]$ $(i=1,2,\ldots,n)$. In what follows, we focus on how to solve Eq. (1.28).

According to the Lagrange multiplier method, the Lagrange function of Eq. (1.28) can be constructed as follows:

$$\hat{L}\left(\overline{x}(\overline{v}),\lambda,\mu\right)=\sum_{S\subseteq N}\left[\left(\sum_{i\in S}x_{Li}(\overline{v})-v_L(S)\right)^2+\left(\sum_{i\in S}x_{Ri}(\overline{v})-v_R(S)\right)^2\right]$$
$$+\lambda\left(\sum_{i=1}^{n}x_{Li}(\overline{v})-v_L(N)\right)+\mu\left(\sum_{i=1}^{n}x_{Ri}(\overline{v})-v_R(N)\right).$$

Then, the optimal solution $\overline{x}^{*\mathrm{E}}(\overline{v})$ of Eq. (1.28) (i.e., the interval-valued least square solution with considering the efficiency) can be obtained through solving the quadratic programming model as follows:

$$\min\left\{\hat{L}\left(\overline{x}(\overline{v}),\lambda,\mu\right)\right\}.$$

The partial derivatives of $\hat{L}\left(\overline{x}(\overline{v}),\lambda,\mu\right)$ with respect to the variables $x_{Lj}(\overline{v}), x_{Rj}(\overline{v})$ $(j\in S\subseteq N)$, λ, and μ are obtained as follows:

$$\frac{\partial\hat{L}\left(\overline{x}(\overline{v}),\lambda,\mu\right)}{\partial x_{Lj}(\overline{v})}=2\sum_{S\subseteq N:j\in S}\left(\sum_{i\in S}x_{Li}(\overline{v})-v_L(S)\right)+\lambda\quad(j=1,2,\ldots,n),$$

$$\frac{\partial\hat{L}\left(\overline{x}(\overline{v}),\lambda,\mu\right)}{\partial\lambda}=\sum_{i=1}^{n}x_{Li}(\overline{v})-v_L(N),$$

$$\frac{\partial\hat{L}\left(\overline{x}(\overline{v}),\lambda,\mu\right)}{\partial x_{Rj}(\overline{v})}=2\sum_{S\subseteq N:j\in S}\left(\sum_{i\in S}x_{Ri}(\overline{v})-v_R(S)\right)+\mu\quad(j=1,2,\ldots,n),$$

and

$$\frac{\partial\hat{L}\left(\overline{x}(\overline{v}),\lambda,\mu\right)}{\partial\mu}=\sum_{i=1}^{n}x_{Ri}(\overline{v})-v_R(N),$$

respectively.

Let the partial derivatives of $\hat{L}\left(\overline{x}(\overline{v}),\lambda,\mu\right)$ with respect to the variables $x_{Lj}(\overline{v})$, $x_{Rj}(\overline{v})$ $(j\in S\subseteq N)$, λ, and μ be equal to 0, respectively. Thus, we have

$$2\sum_{S\subseteq N:j\in S}\left(\sum_{i\in S}x_{Li}^{*\mathrm{E}}(\overline{v})-v_L(S)\right)+\lambda^*=0\quad(j=1,2,\ldots,n),$$

$$\sum_{i=1}^{n}x_{Li}^{*\mathrm{E}}(\overline{v})-v_L(N)=0,$$

$$2\sum_{S\subseteq N:j\in S}\left(\sum_{i\in S}x_{Ri}^{*\mathrm{E}}(\overline{v})-v_R(S)\right)+\mu^*=0\quad(j=1,2,\ldots,n),$$

and

$$\sum_{i=1}^{n} x_{Ri}^{*\mathrm{E}}(\overline{\upsilon}) - \upsilon_R(N) = 0,$$

which infer that

$$\sum_{S\subseteq N:j\in S} \sum_{i\in S} x_{Li}^{*\mathrm{E}}(\overline{\upsilon}) + \frac{\lambda^*}{2} = \sum_{S\subseteq N:j\in S} \upsilon_L(S) \quad (j = 1, 2, \ldots, n), \tag{1.29}$$

$$\sum_{i=1}^{n} x_{Li}^{*\mathrm{E}}(\overline{\upsilon}) = \upsilon_L(N), \tag{1.30}$$

$$\sum_{S\subseteq N:j\in S} \sum_{i\in S} x_{Ri}^{*\mathrm{E}}(\overline{\upsilon}) + \frac{\mu^*}{2} = \sum_{S\subseteq N:j\in S} \upsilon_R(S) \quad (j = 1, 2, \ldots, n), \tag{1.31}$$

and

$$\sum_{i=1}^{n} x_{Ri}^{*\mathrm{E}}(\overline{\upsilon}) = \upsilon_R(N), \tag{1.32}$$

respectively.

Denote $\boldsymbol{x}_L^{*\mathrm{E}}(\overline{\upsilon}) = \left(x_{L1}^{*\mathrm{E}}(\overline{\upsilon}), x_{L2}^{*\mathrm{E}}(\overline{\upsilon}), \ldots, x_{Ln}^{*\mathrm{E}}(\overline{\upsilon})\right)^{\mathrm{T}}$ and $\boldsymbol{e} = (1, 1, \ldots, 1)^{\mathrm{T}}$ which is a n-dimensional vector. Then, Eqs. (1.29) and (1.30) can be rewritten as follows:

$$\boldsymbol{A}\boldsymbol{x}_L^{*\mathrm{E}}(\overline{\upsilon}) + \frac{\lambda^*}{2}\boldsymbol{e} = \boldsymbol{b}_L(\overline{\upsilon}) \tag{1.33}$$

and

$$\boldsymbol{e}^{\mathrm{T}}\boldsymbol{x}_L^{*\mathrm{E}}(\overline{\upsilon}) = \upsilon_L(N), \tag{1.34}$$

respectively, where the vector $\boldsymbol{b}_L(\overline{\upsilon})$ and the matrix $\boldsymbol{A}$ are given by Eqs. (1.15) and (1.17), respectively.

It easily follows from Eq. (1.33) that

$$\boldsymbol{x}_L^{*\mathrm{E}}(\overline{\upsilon}) = \boldsymbol{A}^{-1}\boldsymbol{b}_L(\overline{\upsilon}) - \frac{\lambda^*}{2}\boldsymbol{A}^{-1}\boldsymbol{e},$$

which can be rewritten as follows:

$$\boldsymbol{x}_L^{*\mathrm{E}}(\overline{\upsilon}) = \boldsymbol{x}_L^{*}(\overline{\upsilon}) - \frac{\lambda^*}{2}\boldsymbol{A}^{-1}\boldsymbol{e}, \tag{1.35}$$

where $\boldsymbol{x}_L^*(\overline{v})$ is given by Eq. (1.21). Therefore, in what follows, we need to determine the optimal value λ^*.

It is easily derived from Eqs. (1.34) and (1.35) that

$$\boldsymbol{e}^{\mathrm{T}}\boldsymbol{x}_L^*(\overline{v}) - \frac{\lambda^*}{2}\boldsymbol{e}^{\mathrm{T}}\boldsymbol{A}^{-1}\boldsymbol{e} = v_L(N).$$

Obviously, we have

$$\boldsymbol{e}^{\mathrm{T}}\boldsymbol{x}_L^*(\overline{v}) = \sum_{i=1}^{n} x_{Li}^*(\overline{v})$$

and

$$\boldsymbol{e}^{\mathrm{T}}\boldsymbol{A}^{-1}\boldsymbol{e} = \frac{1}{2^{n-2}} \times \frac{n}{n+1}.$$

Hence, we have

$$\frac{\lambda^*}{2} = 2^{n-2}\frac{n+1}{n}\left(\sum_{i=1}^{n} x_{Li}^*(\overline{v}) - v_L(N)\right). \tag{1.36}$$

Combining with Eq. (1.35), we obtain

$$\begin{aligned} \boldsymbol{x}_L^{*\mathrm{E}}(\overline{v}) &= \boldsymbol{x}_L^*(\overline{v}) - 2^{n-2}\frac{n+1}{n}\left(\sum_{i=1}^{n} x_{Li}^*(\overline{v}) - v_L(N)\right)\boldsymbol{A}^{-1}\boldsymbol{e} \\ &= \boldsymbol{x}_L^*(\overline{v}) - 2^{n-2}\frac{n+1}{n}\left(\sum_{i=1}^{n} x_{Li}^*(\overline{v}) - v_L(N)\right)\left(\frac{1}{2^{n-2}} \times \frac{1}{n+1}\right)\boldsymbol{e} \\ &= \boldsymbol{x}_L^*(\overline{v}) - \frac{1}{n}\left(\sum_{i=1}^{n} x_{Li}^*(\overline{v}) - v_L(N)\right)\boldsymbol{e}, \end{aligned}$$

that is,

$$\boldsymbol{x}_L^{*\mathrm{E}}(\overline{v}) = \boldsymbol{x}_L^*(\overline{v}) + \frac{1}{n}\left(v_L(N) - \sum_{i=1}^{n} x_{Li}^*(\overline{v})\right)\boldsymbol{e}. \tag{1.37}$$

Analogously, denote $\boldsymbol{x}_R^{*\mathrm{E}}(\overline{v}) = \left(x_{R1}^{*\mathrm{E}}(\overline{v}), x_{R2}^{*\mathrm{E}}(\overline{v}), \ldots, x_{Rn}^{*\mathrm{E}}(\overline{v})\right)^{\mathrm{T}}$, then Eqs. (1.31) and (1.32) can be rewritten as follows:

$$\boldsymbol{A}\boldsymbol{x}_R^{*\mathrm{E}}(\overline{v}) + \frac{\mu^*}{2}\boldsymbol{e} = \boldsymbol{b}_R(\overline{v}) \tag{1.38}$$

and

$$e^{\mathrm{T}} x_R^{*\mathrm{E}}(\overline{v}) = v_R(N), \tag{1.39}$$

respectively, where the vector $b_R(\overline{v})$ is given by Eq. (1.16).

It easily follows from Eq. (1.38) that

$$x_R^{*\mathrm{E}}(\overline{v}) = A^{-1} b_R(\overline{v}) - \frac{\mu^*}{2} A^{-1} e,$$

which can be rewritten as follows:

$$x_R^{*\mathrm{E}}(\overline{v}) = x_R^*(\overline{v}) - \frac{\mu^*}{2} A^{-1} e,$$

where $x_R^*(\overline{v})$ is given by Eq. (1.22). Combining with Eq. (1.39), we have

$$e^{\mathrm{T}} x_R^*(\overline{v}) - \frac{\mu^*}{2} e^{\mathrm{T}} A^{-1} e = v_R(N).$$

Obviously, we have

$$e^{\mathrm{T}} x_R^*(\overline{v}) = \sum_{i=1}^{n} x_{Ri}^*(\overline{v}).$$

Hence, we have

$$\frac{\mu^*}{2} = 2^{n-2} \frac{n+1}{n} \left(\sum_{i=1}^{n} x_{Ri}^*(\overline{v}) - v_R(N) \right).$$

Thus, we obtain

$$\begin{aligned} x_R^{*\mathrm{E}}(\overline{v}) &= x_R^*(\overline{v}) - 2^{n-2} \frac{n+1}{n} \left(\sum_{i=1}^{n} x_{Ri}^*(\overline{v}) - v_R(N) \right) A^{-1} e \\ &= x_R^*(\overline{v}) - 2^{n-2} \frac{n+1}{n} \left(\sum_{i=1}^{n} x_{Ri}^*(\overline{v}) - v_R(N) \right) \left(\frac{1}{2^{n-2}} \times \frac{1}{n+1} \right) e \\ &= x_R^*(\overline{v}) - \frac{1}{n} \left(\sum_{i=1}^{n} x_{Ri}^*(\overline{v}) - v_R(N) \right) e, \end{aligned}$$

i.e.,

$$x_R^{*\mathrm{E}}(\overline{v}) = x_R^*(\overline{v}) + \frac{1}{n} \left(v_R(N) - \sum_{i=1}^{n} x_{Ri}^*(\overline{v}) \right) e. \tag{1.40}$$

Therefore, for any interval-valued cooperative game $\overline{\upsilon} \in \overline{G}^n$, we can obtain the interval-valued least square solution $\overline{\boldsymbol{\rho}}^{\text{LSE}}(\overline{\upsilon})$ with considering the efficiency, i.e., $\overline{\boldsymbol{\rho}}^{\text{LSE}}(\overline{\upsilon}) = \overline{\boldsymbol{x}}^{*\text{E}}(\overline{\upsilon})$, whose components are expressed as the intervals $\overline{x}_i^{*\text{E}}(\overline{\upsilon}) = \left[x_{Li}^{*\text{E}}(\overline{\upsilon}), x_{Ri}^{*\text{E}}(\overline{\upsilon})\right]$ $(i = 1, 2, \ldots, n)$, which are given by Eqs. (1.37) and (1.40), respectively.

As stated in Sect. 1.4.1, if all coalitions' values $\overline{\upsilon}(S)$ degenerate to real numbers, then Eqs. (1.37) and (1.40) are identical. Namely, Eqs. (1.37) and (1.40) are applicable to the classical cooperative games.

In a similar way to Sect. 1.4.1, in the following, we can discuss some useful and important properties of the interval-valued least square solutions with considering the efficiency.

Theorem 1.8 *(Existence and Uniqueness) For an arbitrary interval-valued cooperative game* $\overline{\upsilon} \in \overline{G}^n$, *there always exists a unique interval-valued least square solution* $\overline{\boldsymbol{\rho}}^{\text{LSE}}(\overline{\upsilon})$ *with considering the efficiency, which is determined by Eqs. (1.37) and (1.40).*

Proof According to Eqs. (1.37) and (1.40), it is straightforward to prove Theorem 1.8.

Theorem 1.9 *(Efficiency) For an arbitrary interval-valued cooperative game* $\overline{\upsilon} \in \overline{G}^n$, *then its interval-valued least square solution* $\overline{\boldsymbol{\rho}}^{\text{LSE}}(\overline{\upsilon})$ *satisfies the efficiency, i.e.,* $\sum_{i=1}^n \overline{\rho}_i^{\text{LSE}}(\overline{\upsilon}) = \overline{\upsilon}(N)$.

Proof According to Eqs. (1.37) and (1.40), and combining with Definition 1.1, we have

$$\begin{aligned}
\sum_{i=1}^n \overline{\rho}_i^{\text{LSE}}(\overline{\upsilon}) &= \sum_{i=1}^n \left[x_{Li}^*(\overline{\upsilon}) + \frac{1}{n}\left(\upsilon_L(N) - \sum_{i=1}^n x_{Li}^*(\overline{\upsilon})\right), x_{Ri}^*(\overline{\upsilon}) + \frac{1}{n}\left(\upsilon_R(N) - \sum_{i=1}^n x_{Ri}^*(\overline{\upsilon})\right)\right] \\
&= \left[\sum_{i=1}^n x_{Li}^*(\overline{\upsilon}) + \upsilon_L(N) - \sum_{i=1}^n x_{Li}^*(\overline{\upsilon}), \sum_{i=1}^n x_{Ri}^*(\overline{\upsilon}) + \upsilon_R(N) - \sum_{i=1}^n x_{Ri}^*(\overline{\upsilon})\right] \\
&= [\upsilon_L(N), \upsilon_R(N)] \\
&= \overline{\upsilon}(N),
\end{aligned}$$

i.e., $\sum_{i=1}^n \overline{\rho}_i^{\text{LSE}}(\overline{\upsilon}) = \overline{\upsilon}(N)$. Thus, we have completed the proof of Theorem 1.9.

Theorem 1.10 *(Additivity) For any two interval-valued cooperative games* $\overline{\upsilon} \in \overline{G}^n$ *and* $\overline{\nu} \in \overline{G}^n$, *then* $\overline{x}_i^{*\text{E}}(\overline{\upsilon} + \overline{\nu}) = \overline{x}_i^{*\text{E}}(\overline{\upsilon}) + \overline{x}_i^{*\text{E}}(\overline{\nu})$ $(i = 1, 2, \ldots, n)$, *i.e.,* $\overline{\boldsymbol{\rho}}^{\text{LSE}}(\overline{\upsilon} + \overline{\nu}) = \overline{\boldsymbol{\rho}}^{\text{LSE}}(\overline{\upsilon}) + \overline{\boldsymbol{\rho}}^{\text{LSE}}(\overline{\nu})$.

Proof According to Eq. (1.37) and Theorem 1.3, we have

$$\begin{aligned}
\boldsymbol{x}_L^{*\mathrm{E}}(\overline{v}+\overline{\nu}) &= \boldsymbol{x}_L^{*}(\overline{v}+\overline{\nu})+\frac{1}{n}\left[(v_L(N)+\nu_L(N))-\sum_{i=1}^{n}x_{Li}^{*}(\overline{v}+\overline{\nu})\right]\boldsymbol{e}\\
&=\left[\boldsymbol{x}_L^{*}(\overline{v})+\frac{1}{n}\left(v_L(N)-\sum_{i=1}^{n}x_{Li}^{*}(\overline{v})\right)\boldsymbol{e}\right]+\left[\boldsymbol{x}_L^{*}(\overline{\nu})+\frac{1}{n}\left(\nu_L(N)-\sum_{i=1}^{n}x_{Li}^{*}(\overline{\nu})\right)\boldsymbol{e}\right]\\
&=\boldsymbol{x}_L^{*\mathrm{E}}(\overline{v})+\boldsymbol{x}_L^{*\mathrm{E}}(\overline{\nu}),
\end{aligned}$$

i.e., $\boldsymbol{x}_L^{*\mathrm{E}}(\overline{v}+\overline{\nu})=\boldsymbol{x}_L^{*\mathrm{E}}(\overline{v})+\boldsymbol{x}_L^{*\mathrm{E}}(\overline{\nu})$.

Analogously, according to Eq. (1.40) and Theorem 1.3, we can easily prove that $\boldsymbol{x}_R^{*\mathrm{E}}(\overline{v}+\overline{\nu})=\boldsymbol{x}_R^{*\mathrm{E}}(\overline{v})+\boldsymbol{x}_R^{*\mathrm{E}}(\overline{\nu})$. Combining with the case 1 of Definition 1.1, we obtain

$$\overline{x}_i^{*\mathrm{E}}(\overline{v}+\overline{\nu})=\overline{x}_i^{*\mathrm{E}}(\overline{v})+\overline{x}_i^{*\mathrm{E}}(\overline{\nu}) \quad (i=1,2,\ldots,n),$$

i.e.,

$$\overline{\boldsymbol{x}}^{*\mathrm{E}}(\overline{v}+\overline{\nu})=\overline{\boldsymbol{x}}^{*\mathrm{E}}(\overline{v})+\overline{\boldsymbol{x}}^{*\mathrm{E}}(\overline{\nu})$$

or

$$\overline{\boldsymbol{\rho}}^{\mathrm{LSE}}(\overline{v}+\overline{\nu})=\overline{\boldsymbol{\rho}}^{\mathrm{LSE}}(\overline{v})+\overline{\boldsymbol{\rho}}^{\mathrm{LSE}}(\overline{\nu}).$$

Therefore, we have completed the proof of Theorem 1.10.

Theorem 1.11 *(Symmetry) If $i\in N$ and $k\in N(i\neq k)$ are two symmetric players in an interval-valued cooperative game $\overline{v}\in\overline{G}^n$, then $\overline{x}_i^{*\mathrm{E}}(\overline{v})=\overline{x}_k^{*\mathrm{E}}(\overline{v})$, i.e., $\overline{\rho}_i^{\mathrm{LSE}}(\overline{v})=\overline{\rho}_k^{\mathrm{LSE}}(\overline{v})$.*

Proof According to Eqs. (1.37) and (1.40), and combining with Definition 1.3, we can prove Theorem 1.11 in the same way to that of Theorem 1.4 (omitted).

Theorem 1.12 *(Null player) If $i\in N$ is a null player in an interval-valued cooperative game $\overline{v}\in\overline{G}^n$, then $\overline{x}_i^{*\mathrm{E}}(\overline{v})=0$, i.e., $\overline{\rho}_i^{\mathrm{LSE}}(\overline{v})=0$.*

Proof According to Eqs. (1.37) and (1.40), and combining with Definition 1.4, we can prove Theorem 1.12 in the same way to that of Theorem 1.5 (omitted).

Theorem 1.13 *(Dummy player) If $i\in N$ is a dummy player in an interval-valued cooperative game $\overline{v}\in\overline{G}^n$, then $\overline{x}_i^{*\mathrm{E}}(\overline{v})=\overline{v}(i)$, i.e., $\overline{\rho}_i^{\mathrm{LSE}}(\overline{v})=\overline{v}(i)$.*

Proof According to Eqs. (1.37) and (1.40), and combining with Theorem 1.6, we have completed the proof of Theorem 1.13.

Theorem 1.14 *(Anonymity) For any permutation σ on the set N and an interval-valued cooperative game $\overline{v}\in\overline{G}^n$, then $\overline{x}_{\sigma(i)}^{*\mathrm{E}}(\overline{v}^{\sigma})=\overline{x}_i^{*\mathrm{E}}(\overline{v})$, i.e., $\overline{\rho}_{\sigma(i)}^{\mathrm{LSE}}(\overline{v}^{\sigma})=\overline{\rho}_i^{\mathrm{LSE}}(\overline{v})$. Namely, $\overline{\boldsymbol{\rho}}^{\mathrm{LSE}}(\overline{v}^{\sigma})=\boldsymbol{\sigma}^{\#}\left(\overline{\boldsymbol{\rho}}^{\mathrm{LSE}}(\overline{v})\right)$.*

Proof According to Eqs. (1.37) and (1.40), and combining with Theorem 1.7, we have completed the proof of Theorem 1.14.

From Theorems 1.8–1.14, for any interval-valued cooperative game $\overline{\upsilon} \in \overline{G}^n$, there always exists a unique interval-valued least square solution $\overline{\rho}^{\mathrm{LSE}}(\overline{\upsilon})$, which satisfies the efficiency, the additivity, the symmetry, the anonymity, the dummy player property, and the null player property.

1.5 Analysis of Two Examples and Computational Result Comparison

There are many applications of the classical cooperative game theory to real decision problems in finance, management, business, investment, and economics. To illustrate and compare the quadratic programming models and methods proposed in the aforementioned Sect. 1.4 with Han et al.'s method [19], we adopt the same example from [19], which is briefly described as follows. The following Example 1.1 is an interval-valued cooperative game, which is applied to determine optimal allocation strategies of enterprises (or factories).

Example 1.1 Suppose that there are three factories (i.e., players) 1, 2, and 3, who have the ability to produce separately. Denoted the set of players by $N' = \{1, 2, 3\}$. Now, they plan to work together for manufacturing a better product. Due to the incomplete and uncertain information, they cannot precisely forecast their profits (or gains). Generally, they can estimate ranges of their profits. Namely, the profit of a coalition $S \subseteq N'$ of the factories (i.e., players) may be expressed with an interval $\overline{\upsilon}'(S) = [\upsilon_L'(S), \upsilon_R'(S)]$. In this case, the optimal allocation problem of profits for the factories may be regarded as an interval-valued cooperative game $\overline{\upsilon}'$ in which the interval-valued characteristic function is equal to $\overline{\upsilon}'(S)$ for any coalition $S \subseteq N'$. Thus, if they manufacture the product by themselves, then their profits are expressed with the intervals $\overline{\upsilon}'(1) = [0, 2]$, $\overline{\upsilon}'(2) = [1/2, 3/2]$, and $\overline{\upsilon}'(3) = [1, 2]$, respectively. Similarly, if any two factories cooperatively manufacture the product, then their profits are expressed with the intervals $\overline{\upsilon}'(1, 2) = [2, 3]$, $\overline{\upsilon}'(2, 3) = [4, 4]$, and $\overline{\upsilon}'(1, 3) = [3, 4]$, respectively. If all three factories (i.e., the grand coalition N') cooperatively manufacture the product, then the profit is expressed with the interval $\overline{\upsilon}'(1, 2, 3) = [6, 7]$.

1.5.1 Computational Results Obtained by Different Methods and Analysis

In this subsection, the above numerical Example 1.1 is solved by the quadratic programming models and methods proposed in the aforementioned Sect. 1.4 and the method proposed by Han et al. [19]. The computational results are analyzed and compared to show the validity, the applicability, and the superiority of the proposed quadratic programming models and methods.

It is obvious from the above interval-valued coalitions' values $\bar{v}'(S)$ $(S \subseteq N')$ and Eqs. (1.15), (1.16), and (1.20) that

$$\boldsymbol{b}_L\left(\bar{v}'\right) = \left(11, \frac{25}{2}, 14\right)^{\mathrm{T}},$$

$$\boldsymbol{b}_R\left(\bar{v}'\right) = \left(16, \frac{31}{2}, 17\right)^{\mathrm{T}},$$

and

$$\boldsymbol{A}'^{-1} = \begin{pmatrix} \frac{3}{8} & -\frac{1}{8} & -\frac{1}{8} \\ -\frac{1}{8} & \frac{3}{8} & -\frac{1}{8} \\ -\frac{1}{8} & -\frac{1}{8} & \frac{3}{8} \end{pmatrix}.$$

Using the quadratic programming method, i.e., Eqs. (1.21) and (1.22), we can easily obtain

$$\boldsymbol{x}_L^*\left(\bar{v}'\right) = \boldsymbol{A}'^{-1}\boldsymbol{b}_L\left(\bar{v}'\right) = \begin{pmatrix} \frac{3}{8} & -\frac{1}{8} & -\frac{1}{8} \\ -\frac{1}{8} & \frac{3}{8} & -\frac{1}{8} \\ -\frac{1}{8} & -\frac{1}{8} & \frac{3}{8} \end{pmatrix} \begin{pmatrix} 11 \\ \frac{25}{2} \\ 14 \end{pmatrix} = \begin{pmatrix} \frac{13}{16} \\ \frac{25}{16} \\ \frac{37}{16} \end{pmatrix}$$

and

$$\boldsymbol{x}_R^*\left(\bar{v}'\right) = \boldsymbol{A}'^{-1}\boldsymbol{b}_R\left(\bar{v}'\right) = \begin{pmatrix} \frac{3}{8} & -\frac{1}{8} & -\frac{1}{8} \\ -\frac{1}{8} & \frac{3}{8} & -\frac{1}{8} \\ -\frac{1}{8} & -\frac{1}{8} & \frac{3}{8} \end{pmatrix} \begin{pmatrix} 16 \\ \frac{31}{2} \\ 17 \end{pmatrix} = \begin{pmatrix} \frac{31}{16} \\ \frac{27}{16} \\ \frac{39}{16} \end{pmatrix},$$

respectively. Namely,

$$\bar{x}_1^*\left(\bar{v}'\right) = \left[x_{L1}^*\left(\bar{v}'\right), x_{R1}^*\left(\bar{v}'\right)\right] = \left[\frac{13}{16}, \frac{31}{16}\right],$$

$$\bar{x}_2^*\left(\bar{v}'\right) = \left[x_{L2}^*\left(\bar{v}'\right), x_{R2}^*\left(\bar{v}'\right)\right] = \left[\frac{25}{16}, \frac{27}{16}\right],$$

and

$$\bar{x}_3^*\left(\bar{v}'\right) = \left[x_{L3}^*\left(\bar{v}'\right), x_{R3}^*\left(\bar{v}'\right)\right] = \left[\frac{37}{16}, \frac{39}{16}\right].$$

Thus, we obtain the interval-valued least square solution $\bar{\rho}^{\mathrm{LS}}\left(\bar{v}'\right) = \bar{x}^*\left(\bar{v}'\right) = \left(\bar{x}_1^*\left(\bar{v}'\right), \bar{x}_2^*\left(\bar{v}'\right), \bar{x}_3^*\left(\bar{v}'\right)\right)^{\mathrm{T}}$ of the interval-valued cooperative game $\bar{v}' \in \bar{G}^3$, which is the optimal allocation of profits for the cooperative factories (i.e., players) 1, 2, and 3. This optimal allocation may be interpreted as follows: under the cooperation, the factory 1 can obtain at least 13/16 and at most 31/16, i.e., the interval [13/16, 31/16], which is almost greater than the interval $\bar{v}'(1) = [0, 2]$ obtained by itself alone. Analogously, the factory 2 can obtain at least 25/16 and at most 27/16, i.e., the interval [25/16, 27/16], which is obviously greater than the interval $\bar{v}'(2) = [1/2, 3/2]$ obtained by itself alone. The factory 3 can obtain at least 37/16 and at most 39/16, i.e., the interval [37/16, 39/16], which is remarkably greater than the interval $\bar{v}'(3) = [1, 2]$ obtained by itself alone. In other words, the optimal allocations of the factories 2 and 3 satisfy the individual rationality of interval-valued payoff vectors according to the Moore's order relation over intervals given by Eq. (1.4). For the optimal allocation of the factory 1, the interval [13/16, 31/16] is intuitionally better than the interval $\bar{v}'(1) = [0, 2]$ although it does not satisfy the individual rationality of interval-valued payoff vectors according to Eq. (1.4).

Obviously, we have

$$\sum_{i=1}^{3} x_{Li}^*\left(\bar{v}'\right) = \frac{13}{16} + \frac{25}{16} + \frac{37}{16} = \frac{75}{16} \neq 6$$

and

$$\sum_{i=1}^{3} x_{Ri}^*\left(\bar{v}'\right) = \frac{31}{16} + \frac{27}{16} + \frac{39}{16} = \frac{97}{16} \neq 7.$$

That is to say,

$$\sum_{i=1}^{3} \bar{x}_i^*\left(\bar{v}'\right) \neq \bar{v}\left(N'\right),$$

which implies that the interval-valued least square solution $\bar{\rho}^{\mathrm{LS}}(\bar{v}') = (\bar{x}_1^*(\bar{v}'), \bar{x}_2^*(\bar{v}'), \bar{x}_3^*(\bar{v}'))^{\mathrm{T}}$ does not satisfy the efficiency of interval-valued payoff vectors.

If the efficiency condition of interval-valued payoff vectors is considered, then we can use the quadratic programming method with the efficiency to solve the above Example 1.1. More specifically, it easily follows from Eqs. (1.37) and (1.40) that

$$\begin{aligned}
\boldsymbol{x}_L^{*\mathrm{E}}(\bar{v}') &= \boldsymbol{x}_L^*(\bar{v}') + \frac{1}{3}\left(v_L\left(N'\right) - \sum_{i=1}^{3} x_{Li}^*\left(\bar{v}'\right)\right)\boldsymbol{e} \\
&= \left(\frac{13}{16}, \frac{25}{16}, \frac{37}{16}\right)^{\mathrm{T}} + \frac{1}{3}\left(6 - \frac{75}{16}\right)(1,1,1)^{\mathrm{T}} \\
&= \left(\frac{5}{4}, 2, \frac{11}{4}\right)^{\mathrm{T}}
\end{aligned}$$

and

$$\begin{aligned}
\boldsymbol{x}_R^{*\mathrm{E}}(\bar{v}') &= \boldsymbol{x}_R^*(\bar{v}') + \frac{1}{3}\left(v_R\left(N'\right) - \sum_{i=1}^{n} x_{Ri}^*\left(\bar{v}'\right)\right)\boldsymbol{e} \\
&= \left(\frac{31}{16}, \frac{27}{16}, \frac{39}{16}\right)^{\mathrm{T}} + \frac{1}{3}\left(7 - \frac{97}{16}\right)(1,1,1)^{\mathrm{T}} \\
&= \left(\frac{9}{4}, 2, \frac{11}{4}\right)^{\mathrm{T}},
\end{aligned}$$

respectively. Hereby, we have

$$\bar{x}_1^{*\mathrm{E}}\left(\bar{v}'\right) = \left[x_{L1}^{*\mathrm{E}}\left(\bar{v}'\right), x_{R1}^{*\mathrm{E}}\left(\bar{v}'\right)\right] = \left[\frac{5}{4}, \frac{9}{4}\right],$$

$$\bar{x}_2^{*\mathrm{E}}\left(\bar{v}'\right) = \left[x_{L2}^{*\mathrm{E}}\left(\bar{v}'\right), x_{R2}^{*\mathrm{E}}\left(\bar{v}'\right)\right] = [2, 2],$$

and

$$\bar{x}_3^{*\mathrm{E}}\left(\bar{v}'\right) = \left[x_{L3}^{*\mathrm{E}}\left(\bar{v}'\right), x_{R3}^{*\mathrm{E}}\left(\bar{v}'\right)\right] = \left[\frac{11}{4}, \frac{11}{4}\right].$$

Thus, we can obtain the interval-valued least square solution $\bar{\rho}^{\mathrm{LSE}}(\bar{v}') = \bar{x}^{*\mathrm{E}}(\bar{v}') = \left(\left[x_{L1}^{*\mathrm{E}}(\bar{v}'), x_{R1}^{*\mathrm{E}}(\bar{v}')\right], \left[x_{L2}^{*\mathrm{E}}(\bar{v}'), x_{R2}^{*\mathrm{E}}(\bar{v}')\right], \left[x_{L3}^{*\mathrm{E}}(\bar{v}'), x_{R3}^{*\mathrm{E}}(\bar{v}')\right]\right)^{\mathrm{T}}$ with considering the efficiency, which is the optimal allocation of profits of the cooperative factories (i.e., players) 1, 2, and 3 when the efficiency condition is taken into consideration.

This optimal allocation may be interpreted as follows: when the three factories cooperate to form the grand coalition N', the factory 1 can obtain at least 5/4 and at most 9/4, i.e., the interval [5/4, 9/4], which is remarkably greater than the interval $\bar{v}'(1) = [0, 2]$ obtained by itself alone. Analogously, the factory 2 can obtain 2, i.e., the degenerate interval [2, 2], which is obviously greater than the interval $\bar{v}'(2) = [1/2, 3/2]$ obtained by itself alone. The factory 3 can obtain 11/4, i.e., the degenerate interval [11/4, 11/4], which is remarkably greater than the interval $\bar{v}'(3) = [1, 2]$ obtained by itself alone. In other words, the optimal allocations of the factories 1, 2, and 3 satisfy the individual rationality of interval-valued payoff vectors according to the Moore's order relation over intervals given by Eq. (1.4).

Obviously, we have

$$\sum_{i=1}^{3} x_{Li}^{*\mathrm{E}}\left(\bar{v}'\right) = \frac{5}{4} + 2 + \frac{11}{4} = 6$$

and

$$\sum_{i=1}^{3} x_{Ri}^{*\mathrm{E}}\left(\bar{v}'\right) = \frac{9}{4} + 2 + \frac{11}{4} = 7.$$

That is to say,

$$\sum_{i=1}^{3} \bar{x}_i^{*\mathrm{E}}\left(\bar{v}'\right) = \bar{v}'\left(N'\right),$$

which implies that the interval-valued least square solution $\bar{\boldsymbol{\rho}}^{\mathrm{LSE}}\left(\bar{v}'\right) = \left(\bar{x}_1^{*\mathrm{E}}\left(\bar{v}'\right), \bar{x}_2^{*\mathrm{E}}\left(\bar{v}'\right), \bar{x}_3^{*\mathrm{E}}\left(\bar{v}'\right)\right)^{\mathrm{T}}$ satisfies the efficiency of interval-valued payoff vectors as expected.

Therefore, the interval-valued least square solution $\bar{\boldsymbol{\rho}}^{\mathrm{LSE}}\left(\bar{v}'\right)$ is an interval-valued imputation of the interval-valued cooperative game $\bar{v}' \in \bar{G}^3$ according to the definition as stated in Sect. 1.3.2, i.e., $\bar{\boldsymbol{\rho}}^{\mathrm{LSE}}\left(\bar{v}'\right) \in \bar{I}\left(\bar{v}'\right)$.

If Han et al.'s method [19] is used to solve the above numerical Example 1.1, then according to Eq. (4) given by Han et al. [19], the interval-valued Shapley-like value of the factory 1 can be obtained as follows:

$$\overline{\phi}_1^*(\overline{v}') = \sum_{S \subseteq N' \setminus 1} \frac{s!(3-s-1)!}{3!}(\overline{v}'(S \cup 1) - \overline{v}'(S))$$
$$= \frac{0!2!}{3!}\left(\overline{v}'(1) - \overline{v}'(\emptyset)\right) + \frac{1!1!}{3!}\left(\overline{v}'(1,2) - \overline{v}'(2)\right) + \frac{1!1!}{3!}\left(\overline{v}'(1,3) - \overline{v}'(3)\right)$$
$$+ \frac{2!0!}{3!}\left(\overline{v}'(1,2,3) - \overline{v}'(2,3)\right)$$
$$= \frac{0!2!}{3!}([0,2] - [0,0]) + \frac{1!1!}{3!}\left([2,3] - \left[\frac{1}{2}, \frac{3}{2}\right]\right) + \frac{1!1!}{3!}([3,4] - [1,2])$$
$$+ \frac{2!0!}{3!}([6,7] - [4,4])$$
$$= \left[\frac{11}{12}, \frac{31}{12}\right].$$

In the same way, the interval-valued Shapley-like values of the factories 2 and 3 can be obtained as follows:

$$\overline{\phi}_2^*(\overline{v}') = \sum_{S \subseteq N' \setminus 2} \frac{s!(3-s-1)!}{3!}(\overline{v}'(S \cup 2) - \overline{v}'(S))$$
$$= \frac{0!2!}{3!}\left(\overline{v}'(2) - \overline{v}'(\emptyset)\right) + \frac{1!1!}{3!}\left(\overline{v}'(1,2) - \overline{v}'(1)\right) + \frac{1!1!}{3!}\left(\overline{v}'(2,3) - \overline{v}'(3)\right)$$
$$+ \frac{2!0!}{3!}\left(\overline{v}'(1,2,3) - \overline{v}'(1,3)\right)$$
$$= \frac{0!2!}{3!}\left(\left[\frac{1}{2}, \frac{3}{2}\right] - [0,0]\right) + \frac{1!1!}{3!}([2,3] - [0,2]) + \frac{1!1!}{3!}([4,4] - [1,2])$$
$$+ \frac{2!0!}{3!}([6,7] - [3,4])$$
$$= \left[\frac{7}{6}, \frac{17}{6}\right]$$

and

$$\overline{\phi}_3^*(\overline{v}') = \sum_{S \subseteq N' \setminus 3} \frac{s!(3-s-1)!}{3!}(\overline{v}'(S \cup 3) - \overline{v}'(S))$$
$$= \frac{0!2!}{3!}\left(\overline{v}'(3) - \overline{v}'(\emptyset)\right) + \frac{1!1!}{3!}\left(\overline{v}'(2,3) - \overline{v}'(2)\right) + \frac{1!1!}{3!}\left(\overline{v}'(1,3) - \overline{v}'(1)\right)$$
$$+ \frac{2!0!}{3!}\left(\overline{v}'(1,2,3) - \overline{v}'(1,2)\right)$$
$$= \frac{0!2!}{3!}([1,2] - [0,0]) + \frac{1!1!}{3!}\left([4,4] - \left[\frac{1}{2}, \frac{3}{2}\right]\right) + \frac{1!1!}{3!}([3,4] - [0,2])$$
$$+ \frac{2!0!}{3!}([6,7] - [2,3])$$
$$= \left[\frac{23}{12}, \frac{43}{12}\right],$$

respectively.

It easily follows that

$$\sum_{i=1}^{3} \overline{\phi}_i^*\left(\overline{v}'\right) = [4,9] \neq [6,7] = \overline{v}'\left(N'\right),$$

which implies that the interval-valued Shapley-like value (vector) $\overline{\boldsymbol{\Phi}}^*\left(\overline{v}'\right) = \left(\overline{\phi}_1^*\left(\overline{v}'\right), \overline{\phi}_2^*\left(\overline{v}'\right), \overline{\phi}_3^*\left(\overline{v}'\right)\right)^{\mathrm{T}}$ does not satisfy the efficiency. But, the interval-valued Shapley-like value $\overline{\boldsymbol{\Phi}}^*\left(\overline{v}'\right)$ satisfies the individual rationality of interval-valued payoff vectors according to the Moore's order relation over intervals given by Eq. (1.4).

However, if another interval-valued Shapley-like value [19] with satisfying the efficiency is used, then the (classical) median cooperative game v'_{m} and the interval-valued cooperative game $\overline{v}_u$ associated with the above interval-valued cooperative game $\overline{v}'$ need to be firstly defined. More specifically, the median cooperative game v'_{m} has the characteristic function which is defined as

$$v'_{\mathrm{m}}(S) = \frac{v'_L(S) + v'_R(S)}{2}$$

for any coalition $S \subseteq N'$, i.e., $v'_{\mathrm{m}}(1) = 1$, $v'_{\mathrm{m}}(2) = 1$, $v'_{\mathrm{m}}(3) = 3/2$, $v'_{\mathrm{m}}(1,2) = 5/2$, $v'_{\mathrm{m}}(2,3) = 4$, $v'_{\mathrm{m}}(1,3) = 7/2$, and $\overline{v}'_{\mathrm{m}}(1,2,3) = 13/2$. The interval-valued cooperative game $\overline{v}'_u$ has the interval-valued characteristic function which is defined as

$$\overline{v}'_u(S) = \left[-\frac{v'_R(S) - v'_L(S)}{2}, \frac{v'_R(S) - v'_L(S)}{2}\right]$$

for any coalition $S \subseteq N'$, i.e., $\overline{v}'_u(1) = [-1,1]$, $\overline{v}'_u(2) = [-1/2,1/2]$, $\overline{v}'_u(3) = [-1/2,1/2]$, $\overline{v}'_u(1,2) = [-1/2,1/2]$, $\overline{v}'_u(2,3) = [0,0]$, $\overline{v}'_u(1,3) = [-1/2,1/2]$, and $\overline{v}'_u(1,2,3) = [-1/2,1/2]$.

Thus, for the above median cooperative game v'_{m}, we can obtain the Shapley values [10] of the factories 1, 2, and 3 as follows:

$$\begin{aligned}\phi_1^*(v_{\rm m}') &= \sum_{S\subseteq N'\backslash 1} \frac{s!(3-s-1)!}{3!}(v_{\rm m}'(S\cup 1) - v_{\rm m}'(S)) \\ &= \frac{0!2!}{3!}\left(v_{\rm m}'(1) - v_{\rm m}'(\varnothing)\right) + \frac{1!1!}{3!}\left(v_{\rm m}'(1,2) - v_{\rm m}'(2)\right) + \frac{1!1!}{3!}\left(v_{\rm m}'(1,3) - v_{\rm m}'(3)\right) \\ &\quad + \frac{2!0!}{3!}\left(v_{\rm m}'(1,2,3) - v_{\rm m}'(2,3)\right) \\ &= \frac{0!2!}{3!}(1-0) + \frac{1!1!}{3!}\left(\frac{5}{2} - 1\right) + \frac{1!1!}{3!}\left(\frac{7}{2} - \frac{3}{2}\right) + \frac{2!0!}{3!}\left(\frac{13}{2} - 4\right) \\ &= \frac{7}{4},\end{aligned}$$

$$\begin{aligned}\phi_2^*(v_{\rm m}') &= \sum_{S\subseteq N'\backslash 2} \frac{s!(3-s-1)!}{3!}(v_{\rm m}'(S\cup 2) - v_{\rm m}'(S)) \\ &= \frac{0!2!}{3!}\left(v_{\rm m}'(2) - v_{\rm m}'(\varnothing)\right) + \frac{1!1!}{3!}\left(v_{\rm m}'(1,2) - v_{\rm m}'(1)\right) + \frac{1!1!}{3!}\left(v_{\rm m}'(2,3) - v_{\rm m}'(3)\right) \\ &\quad + \frac{2!0!}{3!}\left(v_{\rm m}'(1,2,3) - v_{\rm m}'(1,3)\right) \\ &= \frac{0!2!}{3!}(1-0) + \frac{1!1!}{3!}\left(\frac{5}{2} - 1\right) + \frac{1!1!}{3!}\left(4 - \frac{3}{2}\right) + \frac{2!0!}{3!}\left(\frac{13}{2} - \frac{7}{2}\right) \\ &= 2,\end{aligned}$$

and

$$\begin{aligned}\phi_3^*(v_{\rm m}') &= \sum_{S\subseteq N'\backslash 3} \frac{s!(3-s-1)!}{3!}(v_{\rm m}'(S\cup 3) - v_{\rm m}'(S)) \\ &= \frac{0!2!}{3!}\left(v_{\rm m}'(3) - v_{\rm m}'(\varnothing)\right) + \frac{1!1!}{3!}\left(v_{\rm m}'(2,3) - v_{\rm m}'(2)\right) + \frac{1!1!}{3!}\left(v_{\rm m}'(1,3) - v_{\rm m}'(1)\right) \\ &\quad + \frac{2!0!}{3!}\left(v_{\rm m}'(1,2,3) - v_{\rm m}'(1,2)\right) \\ &= \frac{0!2!}{3!}\left(\frac{3}{2} - 0\right) + \frac{1!1!}{3!}(4-1) + \frac{1!1!}{3!}\left(\frac{7}{2} - 1\right) + \frac{2!0!}{3!}\left(\frac{13}{2} - \frac{5}{2}\right) \\ &= \frac{11}{4},\end{aligned}$$

respectively.

For the above interval-valued cooperative game $\bar{v}_u'$, according to Eq. (4) given by Han et al. [19], the interval-valued Shapley-like values of the factories 1, 2, and 3 can be obtained as follows:

$$
\begin{aligned}
\overline{\phi}_1^*(\overline{v}_u') &= \sum_{S\subseteq N'\setminus 1} \frac{s!(3-s-1)!}{3!}(\overline{v}_u'(S\cup 1)-\overline{v}_u'(S)) \\
&= \frac{0!2!}{3!}\left(\overline{v}_u'(1)-\overline{v}_u'(\varnothing)\right)+\frac{1!1!}{3!}\left(\overline{v}_u'(1,2)-\overline{v}_u'(2)\right)+\frac{1!1!}{3!}\left(\overline{v}_u'(1,3)-\overline{v}_u'(3)\right) \\
&\quad +\frac{2!0!}{3!}\left(\overline{v}_u'(1,2,3)-\overline{v}_u'(2,3)\right) \\
&= \frac{0!2!}{3!}([-1,1]-[0,0])+\frac{1!1!}{3!}\left(\left[-\frac{1}{2},\frac{1}{2}\right]-\left[-\frac{1}{2},\frac{1}{2}\right]\right)+\frac{1!1!}{3!}\left(\left[-\frac{1}{2},\frac{1}{2}\right]-\left[-\frac{1}{2},\frac{1}{2}\right]\right) \\
&\quad +\frac{2!0!}{3!}\left(\left[-\frac{1}{2},\frac{1}{2}\right]-[0,0]\right) \\
&= \left[-\frac{5}{6},\frac{5}{6}\right],
\end{aligned}
$$

$$
\begin{aligned}
\overline{\phi}_2^*(\overline{v}_u') &= \sum_{S\subseteq N'\setminus 2} \frac{s!(3-s-1)!}{3!}(\overline{v}_u'(S\cup 2)-\overline{v}_u'(S)) \\
&= \frac{0!2!}{3!}\left(\overline{v}_u'(2)-\overline{v}_u'(\varnothing)\right)+\frac{1!1!}{3!}\left(\overline{v}_u'(1,2)-\overline{v}_u'(1)\right)+\frac{1!1!}{3!}\left(\overline{v}_u'(2,3)-\overline{v}_u'(3)\right) \\
&\quad +\frac{2!0!}{3!}\left(\overline{v}_u'(1,2,3)-\overline{v}_u'(1,3)\right) \\
&= \frac{0!2!}{3!}\left(\left[-\frac{1}{2},\frac{1}{2}\right]-[0,0]\right)+\frac{1!1!}{3!}\left(\left[-\frac{1}{2},\frac{1}{2}\right]-[-1,1]\right)+\frac{1!1!}{3!}\left([0,0]-\left[-\frac{1}{2},\frac{1}{2}\right]\right) \\
&\quad +\frac{2!0!}{3!}\left(\left[-\frac{1}{2},\frac{1}{2}\right]-\left[-\frac{1}{2},\frac{1}{2}\right]\right) \\
&= \left[-\frac{5}{6},\frac{5}{6}\right],
\end{aligned}
$$

and

$$
\begin{aligned}
\overline{\phi}_3^*(\overline{v}_u') &= \sum_{S\subseteq N'\setminus 3} \frac{s!(3-s-1)!}{3!}(\overline{v}_u'(S\cup 3)-\overline{v}_u'(S)) \\
&= \frac{0!2!}{3!}\left(\overline{v}_u'(3)-\overline{v}_u'(\varnothing)\right)+\frac{1!1!}{3!}\left(\overline{v}_u'(2,3)-\overline{v}_u'(2)\right)+\frac{1!1!}{3!}\left(\overline{v}_u'(1,3)-\overline{v}_u'(1)\right) \\
&\quad +\frac{2!0!}{3!}\left(\overline{v}_u'(1,2,3)-\overline{v}_u'(1,2)\right) \\
&= \frac{0!2!}{3!}\left(\left[-\frac{1}{2},\frac{1}{2}\right]-[0,0]\right)+\frac{1!1!}{3!}\left([0,0]-\left[-\frac{1}{2},\frac{1}{2}\right]\right)+\frac{1!1!}{3!}\left(\left[-\frac{1}{2},\frac{1}{2}\right]-[-1,1]\right) \\
&\quad +\frac{2!0!}{3!}\left(\left[-\frac{1}{2},\frac{1}{2}\right]-\left[-\frac{1}{2},\frac{1}{2}\right]\right) \\
&= \left[-\frac{5}{6},\frac{5}{6}\right],
\end{aligned}
$$

respectively.

Therefore, according to Eq. (9) given by Han et al. [19], the interval-valued Shapley-like value of the factory 1 with satisfying the efficiency can be obtained as follows:

$$\overline{\phi}_1^{\prime *}(\overline{v}^{\prime}) = \phi_1^*(v_{\mathrm{m}}^{\prime}) + \frac{\phi_{R1}^*(\overline{v}_u^{\prime}) - \phi_{L1}^*(\overline{v}_u^{\prime})}{\sum_{i=1}^{3}\left(\phi_{Ri}^*\left(\overline{v}_u^{\prime}\right) - \phi_{Li}^*\left(\overline{v}_u^{\prime}\right)\right)}\left[-\frac{v_R^{\prime}(N^{\prime}) - v_L^{\prime}(N^{\prime})}{2}, \frac{v_R^{\prime}(N^{\prime}) - v_L^{\prime}(N^{\prime})}{2}\right]$$

$$= \frac{7}{4} + \frac{5/6 - (-5/6)}{5/6 - (-5/6) + 5/6 - (-5/6) + 5/6 - (-5/6)}\left[-\frac{7-6}{2}, \frac{7-6}{2}\right]$$

$$= \left[\frac{19}{12}, \frac{23}{12}\right].$$

Analogously, the interval-valued Shapley-like values of the factories 2 and 3 with satisfying the efficiency can be obtained as follows:

$$\overline{\phi}_2^{\prime *}(\overline{v}^{\prime}) = \phi_2^*(v_{\mathrm{m}}^{\prime}) + \frac{\phi_{R2}^*(\overline{v}_u^{\prime}) - \phi_{L2}^*(\overline{v}_u^{\prime})}{\sum_{i=1}^{3}\left(\phi_{Ri}^*\left(\overline{v}_u^{\prime}\right) - \phi_{Li}^*\left(\overline{v}_u^{\prime}\right)\right)}\left[-\frac{v_R^{\prime}(N^{\prime}) - v_L^{\prime}(N^{\prime})}{2}, \frac{v_R^{\prime}(N^{\prime}) - v_L^{\prime}(N^{\prime})}{2}\right]$$

$$= 2 + \frac{5/6 - (-5/6)}{3 \times [5/6 - (-5/6)]}\left[-\frac{7-6}{2}, \frac{7-6}{2}\right]$$

$$= \left[\frac{11}{6}, \frac{13}{6}\right]$$

and

$$\overline{\phi}_3^{\prime *}(\overline{v}^{\prime}) = \phi_3^*(v_{\mathrm{m}}^{\prime}) + \frac{\phi_{R3}^*(\overline{v}_u^{\prime}) - \phi_{L3}^*(\overline{v}_u^{\prime})}{\sum_{i=1}^{3}\left(\phi_{Ri}^*\left(\overline{v}_u^{\prime}\right) - \phi_{Li}^*\left(\overline{v}_u^{\prime}\right)\right)}\left[-\frac{v_R^{\prime}(N^{\prime}) - v_L^{\prime}(N^{\prime})}{2}, \frac{v_R^{\prime}(N^{\prime}) - v_L^{\prime}(N^{\prime})}{2}\right]$$

$$= \frac{11}{4} + \frac{5/6 - (-5/6)}{3 \times [5/6 - (-5/6)]}\left[-\frac{7-6}{2}, \frac{7-6}{2}\right]$$

$$= \left[\frac{31}{12}, \frac{35}{12}\right],$$

respectively. Consequently, the interval-valued Shapley-like value $\overline{\boldsymbol{\Phi}}^{\prime *}(\overline{v}^{\prime})$ can be obtained as follows:

$$\overline{\boldsymbol{\Phi}}^{\prime *}(\overline{v}^{\prime}) = \left(\overline{\phi}_1^{\prime *}(\overline{v}^{\prime}), \overline{\phi}_2^{\prime *}(\overline{v}^{\prime}), \overline{\phi}_3^{\prime *}(\overline{v}^{\prime})\right)^{\mathrm{T}} = \left(\left[\frac{19}{12}, \frac{23}{12}\right], \left[\frac{11}{6}, \frac{13}{6}\right], \left[\frac{31}{12}, \frac{35}{12}\right]\right)^{\mathrm{T}}.$$

Obviously,

$$\sum_{i=1}^{3} \overline{\phi}_i'^{*}\left(\overline{v}'\right) = [6,7] = \overline{v}'\left(N'\right).$$

Also, it is obvious that the interval-valued Shapley-like value $\overline{\boldsymbol{\Phi}}'^{*}(\overline{v}')$ satisfies the individual rationality of interval-valued payoff vectors according to the Moore's order relation over intervals given by Eq. (1.4).

1.5.2 The Comparison Analysis and Conclusion

In order to compare the quadratic programming methods proposed in this chapter with Han et al.'s method [19], another example is given as follows.

Example 1.2 There is an interval-valued cooperative game $\overline{v}'' \in \overline{G}^2$, where the set of players $N'' = \{1,2\}$ and $\overline{v}''(1) = [0.3, 1]$, $\overline{v}''(2) = [2,5]$, and $\overline{v}''(1,2) = [4,6]$.

Using Eq. (4) given by Han et al. [19], we can easily obtain the interval-valued Shapley-like values of the players 1 and 2 as follows:

$$\begin{aligned}
\overline{\phi}_1^{*}(\overline{v}'') &= \sum_{S \subseteq N'' \backslash 1} \frac{s!(2-s-1)!}{2!}(\overline{v}''(S \cup 1) - \overline{v}''(S)) \\
&= \frac{0!1!}{2!}\left(\overline{v}''(1) - \overline{v}''(\varnothing)\right) + \frac{1!0!}{2!}\left(\overline{v}''(1,2) - \overline{v}''(2)\right) \\
&= \frac{0!1!}{2!}([0.3,1] - [0,0]) + \frac{1!0!}{2!}([4,6] - [2,5]) \\
&= [-0.35, 2.5]
\end{aligned}$$

and

$$\begin{aligned}
\overline{\phi}_2^{*}(\overline{v}'') &= \sum_{S \subseteq N'' \backslash 2} \frac{s!(2-s-1)!}{2!}(\overline{v}''(S \cup 2) - \overline{v}''(S)) \\
&= \frac{0!1!}{2!}\left(\overline{v}''(2) - \overline{v}''(\varnothing)\right) + \frac{1!0!}{2!}\left(\overline{v}''(1,2) - \overline{v}''(1)\right) \\
&= \frac{0!1!}{2!}([2,5] - [0,0]) + \frac{1!0!}{2!}([4,6] - [0.3,1]) \\
&= [2.5, 5.35],
\end{aligned}$$

respectively. It is obvious that the lower bound of the interval-valued Shapley-like value $\overline{\phi}_1^{*}(\overline{v}'')$ of the player 1 is a negative number, $\overline{\phi}_1^{*}(\overline{v}'')$ means that the player 1 may get a negative profit (or gain). In other words, the player 1 may get worse if he/she cooperates with the player 2. Clearly, this cooperation between the players 1 and 2 will not happen due to the fact that the profit of the player 1 is not smaller than 0.3 even if he/she does his/her business alone. Therefore, the results obtained through using Eq. (4) given by Han et al. [19] may not be rational. Namely, the interval-valued Shapley-like value proposed by Han et al. [19] may result in irrational results. The main reason for this phenomenon is that the interval-type value $\bar{v}''(1,2) = [4,6]$ of the coalition $\{1,2\}$ overlaps with the interval-type value $\bar{v}''(2) = [2,5]$ of the coalition $\{2\}$. In this case, the subtraction of intervals is not reasonably used.

For the same Example 1.2, if we use the quadratic programming method, then it easily follows from Eqs. (1.15), (1.16), and (1.20) that

$$\boldsymbol{b}_L\left(\bar{v}''\right) = (4.3, 6)^{\mathrm{T}},$$
$$\boldsymbol{b}_R\left(\bar{v}''\right) = (7, 11)^{\mathrm{T}},$$

and

$$\boldsymbol{A}''^{-1} = \begin{pmatrix} \frac{2}{3} & -\frac{1}{3} \\ -\frac{1}{3} & \frac{2}{3} \end{pmatrix}.$$

According to Eqs. (1.21) and (1.22), we can easily obtain

$$\boldsymbol{x}_L^*\left(\bar{v}''\right) = \boldsymbol{A}''^{-1}\boldsymbol{b}_L\left(\bar{v}''\right) = \begin{pmatrix} \frac{2}{3} & -\frac{1}{3} \\ -\frac{1}{3} & \frac{2}{3} \end{pmatrix}\begin{pmatrix} 4.3 \\ 6 \end{pmatrix} = \begin{pmatrix} \frac{26}{30} \\ \frac{77}{30} \end{pmatrix}$$

and

$$\boldsymbol{x}_R^*\left(\bar{v}''\right) = \boldsymbol{A}''^{-1}\boldsymbol{b}_R\left(\bar{v}''\right) = \begin{pmatrix} \frac{2}{3} & -\frac{1}{3} \\ -\frac{1}{3} & \frac{2}{3} \end{pmatrix}\begin{pmatrix} 7 \\ 11 \end{pmatrix} = \begin{pmatrix} 1 \\ 5 \end{pmatrix},$$

respectively. Namely,

$$\bar{x}_1^*\left(\bar{v}''\right) = \left[x_{L1}^*\left(\bar{v}''\right), x_{R1}^*\left(\bar{v}''\right)\right] = \left[\frac{26}{30}, 1\right]$$

and

$$\bar{x}_2^*\left(\bar{v}''\right) = \left[x_{L2}^*\left(\bar{v}''\right), x_{R2}^*\left(\bar{v}''\right)\right] = \left[\frac{77}{30}, 5\right].$$

Obviously, we have

$$x_{L1}^{*}\left(\overline{v}''\right)+x_{L2}^{*}\left(\overline{v}''\right)=\frac{103}{30}=3.4333$$

and

$$x_{R1}^{*}\left(\overline{v}''\right)+x_{R2}^{*}\left(\overline{v}''\right)=6.$$

Thus, the sum of the lower bounds of the interval-valued least square solution $\overline{\rho}^{\mathrm{LS}}(\overline{v}'')=\left(\overline{x}_1^*(\overline{v}''),\overline{x}_2^*(\overline{v}'')\right)^{\mathrm{T}}$ is closer to 4, which is the lower bound of the interval-type value of the grand coalition N''.

In sum, it is not difficult to draw the following conclusions from the aforementioned modeling, solving process and computational results.

1. The quadratic programming methods proposed in this chapter is simpler and more convenient than Han et al.'s method [19] from viewpoint of computational complexity. In the quadratic programming methods, Eqs. (1.21) and (1.22) (or Eqs. (1.37) and (1.40)) can be directly applied to compute the interval-valued least square solutions of interval-valued cooperative games. However, Han et al.'s method [19] is respectively used to compute the interval-valued Shapley-like value of each player.
2. In the quadratic programming methods, the distance is used to measure the differences between interval-valued payoffs and interval-type values of coalitions. Thus, we can effectively avoid the magnification of uncertainty resulted from the subtraction of intervals. However, Han et al.'s method [19] may not overcome this disadvantage. For example, in Example 1.1, the interval lengths of the interval-type values of the coalitions containing the player 2 are not bigger than 1. However, the interval length of the interval-valued Shapley-like value of the player 2 is equal to 10/6, which is remarkably greater than 1.
3. Han et al.'s method [19] may obtain negative interval-valued Shapley-like values of players, which are not rational. For instance, in Example 1.2, $\overline{\phi}_1^*\left(\overline{v}''\right)$ is not a positive interval even if all coalitions' values are positive intervals. However, the quadratic programming methods always assures that the interval-valued least square solutions are positive if all coalitions' values are positive intervals.
4. As stated in Examples 1.1 and 1.2, according to the method proposed by Han et al. [19], the interval lengths of the interval-valued Shapley-like values of players are identical. This conclusion may be unreasonable. In fact, in most management situations, the interval lengths of interval-type values of coalitions may be different. Thus, the ranges (i.e., intervals) of marginal contributions of players are not always identical. Hereby, the interval lengths of interval-valued Shapley-like values of players should be different. As a result, Han et al.'s method [19] cannot always assure the obtained interval-valued Shapley-like values of players are rational.

References

1. Owen G. Game theory. 2nd ed. New York: Academic Press; 1982.
2. Nishizaki I, Sakawa M. Fuzzy and multiobjective games for conflict resolution. Berlin: Springer; 2001.
3. Li D-F. Fuzzy multiobjective many-person decision makings and games. Beijing: National Defense Industry Press; 2003 (in Chinese).
4. Dubois D, Prade H. Fuzzy sets and systems: theory and applications. New York: Academic Press; 1980.
5. Bector CR, Chandra S. Fuzzy mathematical programming and fuzzy matrix games. Berlin: Springer; 2005.
6. Branzei R, Dimitrov D, Tijs S. Shapley-like values for interval bankruptcy games. Econ Bull. 2003;3:1–8.
7. Branzei R, Branzei O, Alparslan Gök SZ, Tijs S. Cooperative interval games: a survey. Cent Eur J Oper Res. 2010;18:397–411.
8. von Neumann J, Morgenstern O. Theory of games and economic behavior. Princeton: Princeton University Press; 1944.
9. Driessen T. Cooperation games: solutions and application. Dordrecht: Kluwer Academic Publisher; 1988.
10. Shapley LS. A value for n-person games. In: Kuhn A, Tucker A, editors. Contributions to the theory of games, Annals of Mathematical Studies, vol. II. Princeton: Princeton University Press; 1953. p. 307–17.
11. Schmeidler D. The nucleolus of a characteristic function game. SIAM J Appl Math. 1969;17:1163–70.
12. Alparslan Gök SZ, Branzei R, Tijs S. The interval Shapley value: an axiomatization. Cent Eur J Oper Res. 2010;18:131–40.
13. Mallozzi L, Scalzo V, Tijs S. Fuzzy interval cooperative games. Fuzzy Set Syst. 2011;165:98–105.
14. Aumann R, Maschler M. Game theoretic analysis of a bankruptcy problem from the Talmud. J Econ Theory. 1985;36:195–213.
15. Moore R. Methods and applications of interval analysis. Philadelphia: SIAM Stud Appl Math; 1979.
16. Alparslan Gök SZ, Miquel S, Tijs S. Cooperation under interval uncertainty. Math Methods Oper Res. 2009;69:99–109.
17. Gillies DB. Some theorems on n-person games. PhD thesis. Princeton: Princeton University Press; 1953.
18. Shapley L. Cores of convex games. Int J Game Theory. 1971;1:11–26.
19. Han W-B, Sun H, Xu G-J. A new approach of cooperative interval games: the interval core and Shapley value revisited. Oper Res Lett. 2012;40:462–8.
20. Mares M. Fuzzy cooperative games. Berlin: Springer; 2001.
21. Branzei R, Alparslan Gök SZ, Branzei O. Cooperation games under interval uncertainty: on the convexity of the interval undominated cores. Cent Eur J Oper Res. 2011;19:523–32.
22. Alparslan Gök SZ, Branzei O, Branzei R, Tijs S. Set-valued solution concepts using interval-type payoffs for interval games. J Math Econ. 2011;47:621–6.
23. Peleg B, Sudhölter P. Introduction to the theory of cooperative games, Series C: game theory, mathematical programming and operations research, vol. 34. 2nd ed. Berlin: Springer; 2007.
24. Branzei R, Dimitrov D, Tijs S. Models in cooperative game theory: crisp, fuzzy, and multi-choice games, Lecture notes in economics and mathematical systems, 556. Berlin: Springer; 2005.
25. Li D-F. Linear programming approach to solve interval-valued matrix games. Omega. 2011; 39(6):655–66.

Chapter 2
Satisfactory Interval-Valued Cores of Interval-Valued Cooperative Games

Abstract The aim of this chapter is to develop an effective nonlinear programming method for computing interval-valued cores of interval-valued cooperative games. In this chapter, we define satisfactory degrees (or ranking indexes) of comparing intervals with the features of inclusion and/or overlap relations and discuss their important properties. Hereby we construct satisfactory crisp equivalent forms of interval-valued inequalities. Based on the concept of interval-valued cores, we derive the auxiliary nonlinear programming models for computing interval-valued cores of interval-valued cooperative games and propose corresponding bisection algorithm, which can always provide global optimal solutions. The developed models and method can provide cooperative chances under the situation of inclusion and/or overlap relations between interval-type coalitions' values in which the Moore's interval ranking method (or order relation between intervals) may not assure that an interval-valued core exists. The proposed method is a generalization of that based on the Moore's interval ranking relation. The feasibility and applicability of the models and method proposed in this chapter are illustrated with a numerical example.

Keywords Interval-valued cooperative game • Core • Interval-valued core • Interval ranking • Mathematical programming • Bisection method

2.1 Introduction

Stated as earlier, cooperative games have many successful applications, especially in enterprise management and economics [1, 2]. However, in real situations, player coalitions' values may be imprecise and vague due to the uncertainty of information and the complexity of players' behavior. As a result, interval-valued cooperative games have been studied [3]. In the foregoing Chap. 1, we proposed the concept of the interval-valued least square solutions of interval-valued cooperative games and discussed their important properties. Hereby we developed fast and effective quadratic programming methods for computing such a kind of interval-valued solutions. The interval-valued least square solution is a single-valued solution concept of interval-valued cooperative games. It is well known that the concept of the core [4], which is a set-valued solution, plays an important role in (classical)

D.-F. Li, *Models and Methods for Interval-Valued Cooperative Games in Economic Management*, DOI 10.1007/978-3-319-28998-4_2

cooperative games [5, 6]. In a very natural way, the core of cooperative games may be extended to the interval-valued core of interval-valued cooperative games. Thereby, Branzei et al. [7] studied the cooperative games under interval uncertainty and the convexity of the interval-valued undominated cores. By introducing the selection of interval-valued cooperative games, Alparslan-Gök et al. [8] and Alparslan-Gök et al. [9] investigated several interval-valued solution concepts of interval-valued cooperative games such as the interval-valued core, the interval-valued dominance core, and stable sets. To study existence of interval-valued cores, they also introduced the notion of Γ-balancedness and extended the Bondareva–Shapley theorem [10] for cooperative games to the interval setting. Han et al. [11] discussed a kind of interval-valued cores through defining a special order relation between intervals. Clearly, all the aforementioned works are conducted on the basis of the traditional interval ranking methods such as the Moore's order relation between intervals [12] (also see Eq. (1.4) in Chap. 1 for a detailed). As stated in the proceeding, these traditional interval ranking methods are relatively strict since they only consider the strict relations including the intersection and being greater whereas they do not consider the inclusion and/or overlap relations between intervals. Additionally, players may accept the inclusion and/or overlap relations between interval-type coalitions' values at some satisfactory degree in practical cooperation. Thus, the main purpose of this chapter is to develop an effective method for computing interval-valued cores of interval-valued cooperative games through introducing the concept of satisfactory degrees (or ranking indexes) of comparing intervals with the feature of the inclusion and/or overlap relations between interval-type coalitions' values.

The rest of this chapter is organized as follows. Section 2.2 gives the concept of satisfactory degrees of comparing intervals, discusses some useful and important properties, and constructs satisfactory crisp equivalent forms of interval-valued inequalities. In Sect. 2.3, we derive the auxiliary nonlinear programming models for computing interval-valued cores of interval-valued cooperative games and propose corresponding bisection method. In Sect. 2.4, a numerical example is used to illustrate the feasibility and applicability of the models and method proposed in this chapter.

2.2 Interval Comparison Satisfactory Degrees and Satisfactory Crisp Equivalent Forms of Interval-Valued Inequalities

The notation of intervals is stated as in Sect. 1.3.1. Namely, $\overline{a} = [a_L, a_R]$ is an interval on the set R of real numbers and $\overline{R}$ is the set of intervals on R.

Alternatively, an interval $\overline{a}$ may be expressed in mean-width (or center-radius) form as $\overline{a} =< m(\overline{a}), w(\overline{a}) >$, where

$$m(\overline{a}) = \frac{a_L + a_R}{2}$$

and

$$w(\overline{a}) = \frac{a_R - a_L}{2}$$

are the mid-point and half-width of the interval $\overline{a} \in \overline{R}$, respectively. Thus, for any intervals $\overline{a} =< m(\overline{a}), w(\overline{a}) >\in \overline{R}$ and $\overline{b} =< m(\overline{b}), w(\overline{b}) >\in \overline{R}$, we can rewrite the addition and the scalar multiplication as follows [13, 14]:

1. $\overline{a} + \overline{b} =< m(\overline{a}) + m(\overline{b}), w(\overline{a}) + w(\overline{b}) >$
2. $\overline{a} - \overline{b} =< m(\overline{a}) - m(\overline{b}), w(\overline{a}) + w(\overline{b}) >$
3. $\gamma\overline{a} =< \gamma m(\overline{a}), |\gamma| w(\overline{a}) >= \begin{cases} < \gamma m(\overline{a}), \gamma w(\overline{a}) > & \text{if } \gamma \geq 0 \\ < \gamma m(\overline{a}), -\gamma w(\overline{a}) > & \text{if } \gamma < 0, \end{cases}$

where $\gamma \in R$ is a real number.

2.2.1 Satisfactory Degrees of Interval Comparison and Properties

The ranking order of intervals is a difficult problem, which has been discussed by some researchers [12, 15, 16]. Moreover, most of the researches about interval-valued cooperative and non cooperative games are conducted on the basis of the interval order relations of Moore [12] and Ishihuchi and Tanaka [15]. Moore [12] believes that $\overline{a} \leq \overline{b}$ if $a_R \leq b_L$, depicted as in the case a of Fig. 1.1. By revising the above Moore's order relation between intervals, Ishihuchi and Tanaka [15] considered that $\overline{a} \leq \overline{b}$ if $a_L \leq b_L$ and $a_R \leq b_R$, depicted as in the cases b and c of Fig. 1.1. The aforementioned two interval ranking methods, which are simply called the Moore's order relation between intervals, are relatively strict in that they only considered the strict relations including the intersection and being greater rather than the inclusion and/or overlap relations between intervals, depicted as in Fig. 1.1. In fact, in terms of the fuzzy set [17, 18], the statement "the interval $\overline{a}$ is not greater than the interval $\overline{b}$" may be regarded as a fuzzy relation between $\overline{a}$ and $\overline{b}$, which is still denoted by $\overline{a} \leq \overline{b}$ for short. Thus, Collins and Hu [19, 20] defined a fuzzy partial order relation between intervals by taking into consideration the inclusion and/or overlap relation between intervals, depicted as in Fig. 2.1.

Comparing Fig. 2.1 with Fig. 1.1, it is obvious that the relations between intervals in the former are more general than those in the latter. In the sequel, we give the concept of satisfactory degrees (or ranking indexes) of intervals' comparison through revising the definition firstly proposed by Collins and Hu [19, 20] (with reference to [21, 22] for a detailed).

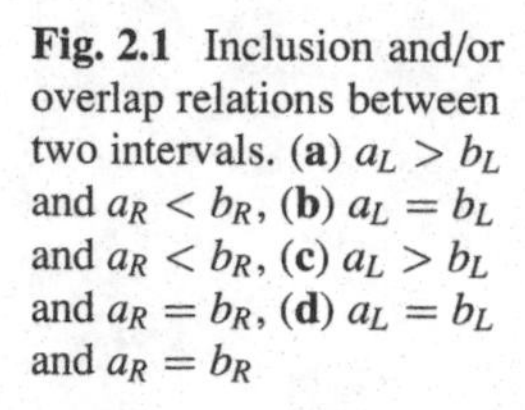

Fig. 2.1 Inclusion and/or overlap relations between two intervals. (**a**) $a_L > b_L$ and $a_R < b_R$, (**b**) $a_L = b_L$ and $a_R < b_R$, (**c**) $a_L > b_L$ and $a_R = b_R$, (**d**) $a_L = b_L$ and $a_R = b_R$

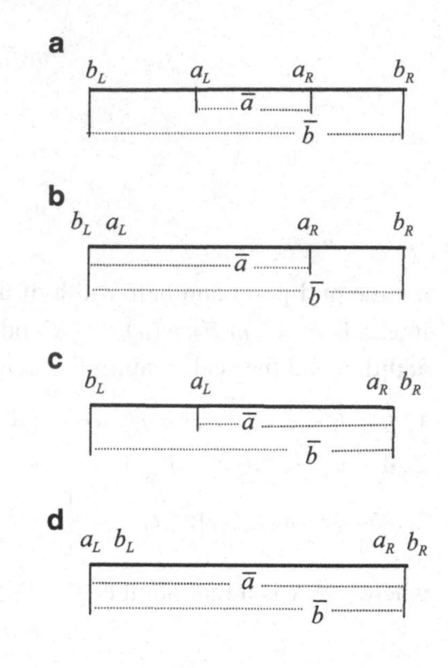

Definition 2.1 Let $\overline{a} = [a_L, a_R] \in \overline{R}$ and $\overline{b} = [b_L, b_R] \in \overline{R}$ be intervals. The premise "$\overline{a} \leq \overline{b}$" is regarded as a fuzzy set, whose membership function is defined as follows:

$$\varphi(\overline{a} \leq \overline{b}) = \begin{cases} 1 & \text{if } a_R < b_L \\ 1^- & \text{if } a_L < b_L \leq a_R < b_R \\ \dfrac{b_R - a_R}{2(w(\overline{b}) - w(\overline{a}))} & \text{if } b_L \leq a_L \leq a_R \leq b_R \text{ and } w(\overline{b}) > w(\overline{a}) \\ 0.5 & \text{if } w(\overline{a}) = w(\overline{b}) \text{ and } a_L = b_L, \end{cases} \tag{2.1}$$

where "1^-" is a fuzzy number of "being less than 1," which indicates that the interval $\overline{a}$ is weakly not greater than the interval $\overline{b}$. The fuzzy number "1^-" may be adequately chosen according to management situations [1, 18, 23, 24].

"$\overline{a} \leq \overline{b}$" is an interval order relation between $\overline{a}$ and $\overline{b}$, which may be regarded as a generalization of the order relation "$a \leq b$" in the set R of real numbers and has the linguistic interpretation "the interval $\overline{a}$ is essentially not greater than the interval $\overline{b}$." Analogously, we can explain "$\overline{a} \geq \overline{b}$" and "$\overline{a} = \overline{b}$."

Obviously, $0 \leq \varphi(\overline{a} \leq \overline{b}) \leq 1$. Thus, $\varphi(\overline{a} \leq \overline{b})$ may be interpreted as the satisfactory degree (or ranking index) of the premise (or order relation) $\overline{a} \leq \overline{b}$. If $\varphi(\overline{a} \leq \overline{b}) = 0$, then the premise $\overline{a} \leq \overline{b}$ is not accepted by the players. If

$\varphi(\bar{a} \leq \bar{b}) = 1$, then the players are absolutely satisfied with the premise $\bar{a} \leq \bar{b}$. That is to say, the players believe that the premise $\bar{a} \leq \bar{b}$ is absolutely true. If $\varphi(\bar{a} \leq \bar{b}) \in (0, 1)$, then the players accept the premise $\bar{a} \leq \bar{b}$ with different satisfactory degrees between 0 and 1.

In addition, it is obvious from Definition 2.1 that the satisfactory degree of the premise (or order relation) $\bar{a} \leq \bar{b}$ is equal to 0.5 when two intervals $\bar{a}$ and $\bar{b}$ entirely overlap, depicted as in the case d of Fig. 2.1. Moreover, if two intervals degenerate to an identical real number, then the satisfactory degree of the premise $\bar{a} \leq \bar{b}$ is also equal to 0.5. Apparently, the satisfactory degree of the premise $\bar{a} \leq \bar{b}$ is equal to 1 if $a_R < b_L$, depicted as in the case a of Fig. 1.1. Therefore, the interval order relation given by Definition 2.1 includes the Moore's order relation between intervals. If two intervals have the inclusion relation with $w(\bar{b}) > w(\bar{a})$, depicted as in the cases a–c of Fig. 2.1, then we can easily know that the satisfactory degree of the premise $\bar{a} \leq \bar{b}$ is between 0 and 1 according to Eq. (2.1). Due to the condition $w(\bar{b}) > w(\bar{a})$, both the intervals a and $\bar{b}$ cannot be reduced to the real numbers at the same time even if $w(\bar{a})$ is equal to 0 or approaches to 0, therefore the satisfactory degree of the premise $\bar{a} \leq \bar{b}$ is also between 0 and 1.

Analogously, we can define the following premise "$\bar{a} \geq \bar{b}$," which indicates the statement "the interval $\bar{a}$ is not smaller than the interval $\bar{b}$."

Definition 2.2 Let $\bar{a} = [a_L, a_R] \in \overline{R}$ and $\bar{b} = [b_L, b_R] \in \overline{R}$ be intervals. The premise "$\bar{a} \geq \bar{b}$" is regarded as a fuzzy set, whose membership function is defined as

$$\varphi(\bar{a} \geq \bar{b}) = 1 - \varphi(\bar{a} \leq \bar{b}),$$

i.e.,

$$\varphi(\bar{a} \geq \bar{b}) = \begin{cases} 0 & \text{if } a_R < b_L \\ 0^+ & \text{if } a_L < b_L \leq a_R < b_R \\ \dfrac{a_L - b_L}{2(w(\bar{b}) - w(\bar{a}))} & \text{if } b_L \leq a_L \leq a_R \leq b_R \text{ and } w(\bar{b}) > w(\bar{a}) \\ 0.5 & \text{if } w(\bar{a}) = w(\bar{b}) \text{ and } a_L = b_L, \end{cases} \tag{2.2}$$

where "0^+" is a fuzzy number of "being greater than 0," which linguistically indicates that the interval $\bar{a}$ is weakly not smaller than the interval $\bar{b}$.

Thus, the interval-valued equality relation "=" can be defined as follows: $\bar{a} = \bar{b}$ if and only if $\bar{a} \geq \bar{b}$ and $\bar{a} \leq \bar{b}$. Alternatively, it is derived from Definitions 2.1 and 2.2 that $\bar{a} = \bar{b}$ is equivalent to both $a_L = b_L$ and $a_R = b_R$. Linguistically, "$\bar{a} = \bar{b}$" may be interpreted as "the interval $\bar{a}$ is equal to the interval $\bar{b}$" in the sense of

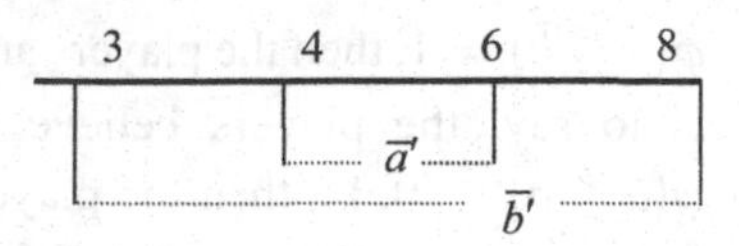

Fig. 2.2 Inclusion relation between the intervals $\overline{a}'$ and $\overline{b}'$

Definitions 2.1 and 2.2. Analogously, $\overline{a} > \overline{b}$ if and only if $\overline{a} \geq \overline{b}$ and $\overline{a} \neq \overline{b}$. $\overline{a} < \overline{b}$ if and only if $\overline{a} \leq \overline{b}$ and $\overline{a} \neq \overline{b}$.

In the sequent, the above fuzzy ranking index φ is often called the satisfactory degree. It is easy to prove that φ is continuous except a single special case, i.e., $a_L = b_L$ and $w(\overline{a}) = w(\overline{b})$.

Example 2.1 Let us consider two intervals $\overline{a}' = [4, 6]$ and $\overline{b}' = [3, 8]$, depicted as in Fig. 2.2.

According to the Moore's order relation between intervals (i.e., Eq. (1.4)), we cannot compare the intervals $\overline{a}'$ and $\overline{b}'$ or rank the order of the intervals $\overline{a}'$ and $\overline{b}'$. However, according to Eq. (2.1), we can obtain

$$\varphi\left(\overline{a}' \leq \overline{b}'\right) = \frac{b_R' - a_R'}{2\left(w(\overline{b}') - w(\overline{a}'')\right)}$$

$$= \frac{8 - 6}{(8 - 3) - (6 - 4)}$$

$$= \frac{2}{3}.$$

Thus, the satisfactory degree of $\overline{a}' \leq \overline{b}'$ is 2/3 whereas the satisfactory degree of $\overline{a}' \geq \overline{b}'$ is 1/3 according to Definition 2.2. In other words, the statement "$\overline{a}' \leq \overline{b}'$" is true with the possibility 2/3.

Moreover, it is easily derived from Definitions 2.1 and 2.2 that there are some useful and important properties [19], which can be summarized as in Theorem 2.1 as follows.

Theorem 2.1 *For any intervals $\overline{a} \in \overline{R}$, $\overline{b} \in \overline{R}$, and $\overline{c} \in \overline{R}$, then*

1. $0 \leq \varphi(\overline{a} \leq \overline{b}) \leq 1$
2. $\varphi(\overline{a} \leq \overline{a}) = 0.5$
3. $\varphi(\overline{a} \leq \overline{b}) + \varphi(\overline{a} \geq \overline{b}) = 1$
4. *if* $\varphi(\overline{a} \leq \overline{b}) \geq 0.5$ *and* $\varphi(\overline{b} \leq \overline{c}) \geq 0.5$, *then* $\varphi(\overline{a} \leq \overline{c}) \geq 0.5$; *or if* $\varphi(\overline{a} \leq \overline{b}) \leq 0.5$ *and* $\varphi(\overline{b} \leq \overline{c}) \leq 0.5$, *then* $\varphi(\overline{a} \leq \overline{c}) \leq 0.5$.

Proof. According to Definitions 2.1 and 2.2, we can easily prove that the conclusions of Theorem 2.1 are valid (omitted).

Thus, Definitions 2.1 and 2.2 may provide quantitative methods to determine the exact satisfactory degree for ranking/comparing two intervals. In the proceeding subsection, the satisfactory degree φ is used to define satisfactory crisp equivalent forms of interval-valued inequalities.

2.2.2 *Satisfactory Crisp Equivalent Forms of Interval-Valued Inequalities*

According to the concept of the satisfactory degrees given above, we can establish the following satisfactory crisp equivalent forms of interval-valued inequality constraints, which will be used to construct auxiliary nonlinear programming models for computing interval-valued cores of interval-valued cooperative games.

Let $\beta \in [0, 1]$ denote the satisfactory degree of the interval-valued inequality constraint $\overline{a} \leq \overline{b}$ which may be satisfied. Then, for the situation in which the intervals $\overline{a}$ and $\overline{b}$ satisfy the following two constraint conditions: $b_L \leq a_L \leq a_R \leq b_R$ and $w(\overline{b}) > w(\overline{a})$, then according to Definition 2.1, the satisfactory crisp equivalent form of the interval-valued inequality $\overline{a} \leq \overline{b}$ is defined as follows:

$$\begin{cases} a_L \geq b_L \\ a_R \leq b_R \\ \varphi(\overline{a} \leq \overline{b}) \geq \beta, \end{cases}$$

which can be further written as the following system of inequalities:

$$\begin{cases} a_L \geq b_L \\ a_R \leq b_R \\ \dfrac{b_R - a_R}{2(w(\overline{b}) - w(\overline{a}))} \geq \beta. \end{cases} \tag{2.3}$$

It is easy to see from Eq. (2.3) that $w(\overline{b}) > w(\overline{a})$ due to $b_R \geq a_R$ and $\beta \in [0, 1]$.

Analogously, for the situation in which the intervals $\overline{a}$ and $\overline{b}$ satisfy the following constraint condition: $a_R < b_L$, then according to Definition 2.1, the satisfactory crisp equivalent form of the interval-valued inequality $\overline{a} \leq \overline{b}$ is defined as follows:

$$a_R < b_L, \tag{2.4}$$

where $\varphi(\overline{a} \leq \overline{b}) = 1$.

For the situation in which the intervals $\overline{a}$ and $\overline{b}$ satisfy the following constraint condition: $a_L < b_L \leq a_R < b_R$, then according to Definition 2.1, the satisfactory crisp equivalent form of the interval-valued inequality $\overline{a} \leq \overline{b}$ is defined as follows:

$$\begin{cases} a_L < b_L \\ b_L \leq a_R \\ a_R < b_R, \end{cases} \tag{2.5}$$

where $\varphi(\overline{a} \leq \overline{b}) = 1^-$.

For the situation in which the intervals $\overline{a}$ and $\overline{b}$ satisfy the following two constraint conditions: $a_L = b_L$ and $w(\overline{a}) = w(\overline{b})$, then according to Definition 2.1, the satisfactory crisp equivalent form of the interval-valued inequality $\overline{a} \leq \overline{b}$ is defined as follows:

$$\begin{cases} a_L = b_L \\ w(\overline{a}) = w(\overline{b}), \end{cases}$$

where $\varphi(\overline{a} \leq \overline{b}) = 0.5$. The above system of equalities can be further written as the following system of equalities:

$$\begin{cases} a_L = b_L \\ a_R - a_L = b_R - b_L. \end{cases} \tag{2.6}$$

In the same way, we can derive the satisfactory crisp equivalent form of the interval-valued inequality $\overline{a} \geq \overline{b}$. More specifically, for the situation in which the intervals $\overline{a}$ and $\overline{b}$ satisfy the following two constraint conditions: $b_L \leq a_L \leq a_R \leq b_R$ and $w(\overline{b}) > w(\overline{a})$, then according to Definition 2.2, the satisfactory crisp equivalent form of the interval-valued inequality $\overline{a} \geq \overline{b}$ is defined as follows:

$$\begin{cases} a_L \geq b_L \\ a_R \leq b_R \\ \varphi(\overline{a} \geq \overline{b}) \geq \beta, \end{cases}$$

which can be further written as the following system of inequalities:

$$\begin{cases} a_L \geq b_L \\ a_R \leq b_R \\ \dfrac{a_L - b_L}{2\left(w(\overline{b}) - w(\overline{a})\right)} \geq \beta. \end{cases} \tag{2.7}$$

Similarly, for the situation in which the intervals $\overline{a}$ and $\overline{b}$ satisfy the following constraint condition: $a_R < b_L$, then according to Definition 2.2, the satisfactory crisp equivalent form of the interval-valued inequality $\overline{a} \geq \overline{b}$ is defined as follows:

$$a_R < b_L, \tag{2.8}$$

where $\varphi(\overline{a} \geq \overline{b}) = 0$.

For the situation in which the intervals $\overline{a}$ and $\overline{b}$ satisfy the following constraint condition: $a_L < b_L \leq a_R < b_R$, then according to Definition 2.2, the satisfactory crisp equivalent form of the interval-valued inequality $\overline{a} \geq \overline{b}$ is defined as follows:

$$\begin{cases} a_L < b_L \\ b_L \leq a_R \\ a_R < b_R, \end{cases} \tag{2.9}$$

where $\varphi(\overline{a} \geq \overline{b}) = 0^+$.

For the situation in which the intervals $\overline{a}$ and $\overline{b}$ satisfy the following two constraint conditions: $a_L = b_L$ and $w(\overline{a}) = w(\overline{b})$, then according to Definition 2.2, the satisfactory crisp equivalent form of the interval-valued inequality $\overline{a} \geq \overline{b}$ is defined as follows:

$$\begin{cases} a_L = b_L \\ w(\overline{a}) = w(\overline{b}), \end{cases}$$

which can be further written as the following system of equalities:

$$\begin{cases} a_L = b_L \\ a_R - a_L = b_R - b_L, \end{cases} \tag{2.10}$$

where $\varphi(\overline{a} \geq \overline{b}) = 0.5$.

2.3 Nonlinear Programming Models and Method for Interval-Valued Cores of Interval-Valued Cooperative Games

In this section, let us continue to consider how to solve interval-valued cooperative games $\overline{v} \in \overline{G}^n$, which are stated in Sect. 1.3.2.

2.3.1 The Concept of Interval-Valued Cores of Interval-Valued Cooperative Games

Stated as earlier, for any interval-valued cooperative game $\overline{v} \in \overline{G}^n$, its interval-valued imputation set $\overline{I}(\overline{v})$ usually may be very large. As a result, in a parallel way to the concept of cores of cooperative games [1, 2, 4, 25], we may give the concept of interval-valued cores of interval-valued cooperative games. More precisely, the interval-valued core of an arbitrary interval-valued cooperative game $\overline{v} \in \overline{G}^n$, denoted by $\overline{C}(\overline{v})$, is defined as follows:

$$\overline{C}(\overline{v}) = \left\{ \overline{\boldsymbol{x}}(\overline{v}) \in \overline{I}(\overline{v}) \middle| \sum_{i \in S} \overline{x}_i(\overline{v}) \geq \overline{v}(S) \text{ for all } S \subset N \right\}, \tag{2.11}$$

where $\overline{\boldsymbol{x}}(\overline{v}) = (\overline{x}_1(\overline{v}), \overline{x}_2(\overline{v}), \ldots, \overline{x}_n(\overline{v}))^{\mathrm{T}}$ and $\overline{x}_i(\overline{v}) = [x_{Li}(\overline{v}), x_{Ri}(\overline{v})]$ $(i = 1, 2, \ldots, n)$ are stated in Sect. 1.3.2.

Obviously, for an inessential interval-valued cooperative game $\overline{v} \in \overline{G}^n$, whose interval-valued characteristic function $\overline{v}$ is defined as

$$\overline{v}(S) = \sum_{i \in S} \overline{v}(i)$$

for any coalition $S \subseteq N$, i.e., the inessential interval-valued cooperative game $\overline{v}$ is additive, hence its interval-valued core $\overline{C}(\overline{v})$ has a unique element, i.e.,

$$\overline{C}(\overline{v}) = \left\{ (\overline{v}(1), \overline{v}(2), \ldots, \overline{v}(n))^{\mathrm{T}} \right\} = \overline{I}(\overline{v}).$$

Clearly, an inessential interval-valued cooperative game is trivial from a game-theoretic point of view. That is to say, if every player $i \in N$ demands at least $\overline{v}(i)$, then the allocation (or distribution) of the grand coalition $\overline{v}(N)$ can be uniquely determined.

Conversely, for an essential interval-valued cooperative game $\overline{v} \in \overline{G}^n$, which is not additive, then its interval-valued core $\overline{C}(\overline{v})$ may have lots of elements if $\overline{C}(\overline{v})$ is not empty, i.e., $\left|\overline{C}(\overline{v})\right| \geq 1$.

It is obvious from Eq. (2.11) that the interval-valued core $\overline{C}(\overline{v})$ of an interval-valued cooperative game $\overline{v} \in \overline{G}^n$ can be obtained through solving the system of linear interval-valued inequalities as follows:

$$\begin{cases} \sum_{i \in S} \overline{x}_i(\overline{v}) \geq \overline{v}(S) \text{ for all } S \subset N \\ \sum_{i=1}^{n} \overline{x}_i(\overline{v}) = \overline{v}(N), \end{cases} \tag{2.12}$$

where $\overline{x}_i(\overline{v}) = [x_{Li}(\overline{v}), x_{Ri}(\overline{v})]$ $(i = 1, 2, \ldots, n)$ are interval-valued variables.

Example 2.2 There are an investor and two IT technologists, numbered by 1, 2, and 3, respectively. The investor 1 has a fund and looks for a technology patent to invest production. The IT technologists 2 and 3 have a similar IT patent and look for a fund to invest production. Due to the uncertainty of the market demand, it seems to be suitable for expressing the profits with intervals. Thus, this problem may be regarded as a three-person interval-valued cooperative game $\overline{v}'' \in \overline{G}^3$, where the investor 1 and the IT technologists 2 and 3 are regarded as the players 1, 2, and 3, respectively; the grand coalition is $N'' = \{1, 2, 3\}$; its interval-valued characteristic function (i.e., profit function) $\overline{v}''$ is defined as follows: $\overline{v}''(1,2) = \overline{v}''(1,3) = \overline{v}''(N'') = [291, 306]$ and $\overline{v}''(S) = 0$ for any other coalitions $S \subset N''$.

Obviously, according to Eq. (2.12), we can easily obtain the interval-valued core of the above interval-valued cooperative game $\overline{v}''$ as follows:

$$\overline{C}\left(\overline{v}''\right) = \left\{([291, 306], 0, 0)^{\mathrm{T}}\right\},$$

which means that there is a unique element (i.e., allocation). In this case, the player 1 (i.e., investor) gets the total profit [291, 306] whereas the players (i.e., technologists) 2 and 3 get nothing from the cooperative production. Clearly, this allocation of the grand coalition $\overline{v}''(N'')$ seems to be irrational due to the following two aspects. On the one hand, the interval-valued characteristic function $\overline{v}''$ is too simple and special to reflect the real situation. For example, it is obvious that

$$\overline{v}''(1,2,3) = \overline{v}''(1,2) + \overline{v}''(3)$$

and

$$\overline{v}''(1,2,3) = \overline{v}''(1,3) + \overline{v}''(2)$$

due to $\overline{v}''(3) = 0$ and $\overline{v}''(2) = 0$. More importantly, on the other hand, the interval-valued cooperative game $\overline{v}''$ is not convex in that

$$\begin{aligned}\overline{v}''(\{1,2\} \cup \{1,3\}) + \overline{v}''(\{1,2\} \cap \{1,3\}) &= \overline{v}''(1,2,3) + \overline{v}''(1) \\ &= [291, 306] < \overline{v}''(1,2) + \overline{v}''(1,3) \\ &= [582, 612].\end{aligned}$$

Moreover, stated as above, the interval-valued cooperative game $\overline{v}''$ is not strictly superadditive in that

$$\overline{v}''(1,2,3) = \overline{v}''(1,2) + \overline{v}''(3)$$

and

$$\overline{v}''(1,2,3) = \overline{v}''(1,3) + \overline{v}''(2).$$

In fact, the interval-valued cooperative game $\overline{v}''$ is a big boss interval-valued cooperative game. For details regarding the solutions and special properties for big boss interval-valued games, we refer the reader to Branzei et al. [26] and Alparslan-Gök et al. [27].

From the above discussion, the interval-valued core of the interval-valued cooperative game $\overline{v}''$ given in Example 2.2 is very easily obtained by simply observation. In many economic management situations, however, it is very difficult to compute interval-valued cores of interval-valued cooperative games. Particularly, if the interval-valued inequality constraints

$$\sum_{i \in S} \overline{x}_i(\overline{v}) \geq \overline{v}(S) \quad (S \subset N)$$

in Eq. (2.12) are made in the sense of the Moore's order relation between intervals (i.e., Eq. (1.4)), then it is rather possible that there exists no feasible solution to Eq. (2.12) and hereby the interval-valued core $\overline{C}(\overline{v})$ of the interval-valued cooperative game $\overline{v} \in \overline{G}^n$ is empty. Although Alparslan-Gök et al. [8] proved that the interval-valued core of an interval-valued cooperative game is non-empty if and only if the interval-valued cooperative game is the Γ-balanced. But, as stated earlier, the Moore's order relation between intervals is very strict and hereby only a special kind of interval-valued cooperative games has non-empty interval-valued cores.

In the next subsection, we focus on developing a satisfactory-degree-based nonlinear programming method for computing interval-valued cores of interval-valued cooperative games.

2.3.2 Nonlinear Programming Models for Interval-Valued Cores of Interval-Valued Cooperative Games

In this section, we mainly apply Definitions 2.1 and 2.2 (i.e., Eqs. (2.7)–(2.10)) to establish the auxiliary nonlinear programming models for Eq. (2.12).

More specifically, for any coalition $S \subset N$, let

$$\beta_S(\overline{v}) = \varphi\left(\sum_{i \in S} \overline{x}_i(\overline{v}) \geq \overline{v}(S)\right)$$

denote the satisfactory degree of the interval-valued inequality $\sum_{i \in S} \overline{x}_i(\overline{v}) \geq \overline{v}(S)$ which may be satisfied.

For the situation in which the intervals $\sum_{i\in S}\overline{x}_i(\overline{v}) = \left[\sum_{i\in S}x_{Li}(\overline{v}), \sum_{i\in S}x_{Ri}(\overline{v})\right]$ and $\overline{v}(S) = [v_L(S), v_R(S)]\,(S \subset N)$ satisfy the constraints as follows:

$$v_L(S) \le \sum_{i\in S} x_{Li}(\overline{v}) \le \sum_{i\in S} x_{Ri}(\overline{v}) \le v_R(S)$$

and

$$w(\overline{v}(S)) > w\left(\sum_{i\in S}\overline{x}_i\right),$$

then according to Eq. (2.7), the satisfactory crisp equivalent mathematical programming model for Eq. (2.12) can be constructed as follows:

$$\max\left\{\min_{S\subset N}\{\beta_S(\overline{v})\}\right\}$$
$$\text{s.t.}\begin{cases}\sum_{i\in S} x_{Li}(\overline{v}) \ge v_L(S) \quad (S\subset N)\\ \sum_{i\in S} x_{Ri}(\overline{v}) \le v_R(S) \quad (S\subset N)\\ \beta_S(\overline{v}) = \varphi\left(\sum_{i\in S}\overline{x}_i \ge \overline{v}(S)\right) \quad (S\subset N)\\ \sum_{i\in N} x_{Ri}(\overline{v}) = v_R(N)\\ \sum_{i\in N} x_{Li}(\overline{v}) = v_L(N)\\ x_{Ri}(\overline{v}) \ge x_{Li}(\overline{v}) \quad (i = 1, 2, \ldots, n).\end{cases} \tag{2.13}$$

Let

$$\beta(\overline{v}) = \min_{S\subset N}\{\beta_S(\overline{v})\}.$$

Then, obviously, $0 \le \beta(\overline{v}) \le 1$. Thereby, according to Definition 2.2 (i.e., Eq. (2.2)), Eq. (2.13) can be rewritten as the following nonlinear programming model:

$$\max\{\beta(\overline{v})\}$$
$$\text{s.t.}\begin{cases}\sum_{i\in S} x_{Li}(\overline{v}) \ge v_L(S) \quad (S\subset N)\\ \sum_{i\in S} x_{Ri}(\overline{v}) \le v_R(S) \quad (S\subset N)\\ \dfrac{\sum_{i\in S} x_{Li}(\overline{v}) - v_L(S)}{(v_R(S)-v_L(S)) - \left(\sum_{i\in S} x_{Ri}(\overline{v}) - \sum_{i\in S} x_{Li}(\overline{v})\right)} \ge \beta(\overline{v}) \quad (S\subset N)\\ \sum_{i\in N} x_{Ri}(\overline{v}) = v_R(N)\\ \sum_{i\in N} x_{Li}(\overline{v}) = v_L(N)\\ 0\le \beta(\overline{v})\le 1\\ x_{Ri}(\overline{v}) \ge x_{Li}(\overline{v}) \quad (i=1,2,\ldots,n),\end{cases}$$

which can be rewritten as the following nonlinear programming model:

$$\max\{\beta(\overline{v})\}$$
$$\text{s.t.}\begin{cases}\sum_{i\in S} x_{Li}(\overline{v}) \ge v_L(S) \quad (S\subset N)\\ \sum_{i\in S} x_{Ri}(\overline{v}) \le v_R(S) \quad (S\subset N)\\ (1-\beta(\overline{v}))\sum_{i\in S} x_{Li}(\overline{v}) + \beta(\overline{v})\sum_{i\in S} x_{Ri}(\overline{v}) \ge (1-\beta(\overline{v}))v_L(S) + \beta(\overline{v})v_R(S) \quad (S\subset N)\\ \sum_{i\in N} x_{Ri}(\overline{v}) = v_R(N)\\ \sum_{i\in N} x_{Li}(\overline{v}) = v_L(N)\\ 0\le \beta(\overline{v})\le 1\\ x_{Ri}(\overline{v}) \ge x_{Li}(\overline{v}) \quad (i=1,2,\ldots,n),\end{cases} \tag{2.14}$$

where $x_{Ri}(\overline{v}), x_{Li}(\overline{v})$ $(i=1,2,\ldots,n)$, and $\beta(\overline{v})$ are decision variables, which need to be determined.

Analogously, for the situation in which the intervals $\sum_{i\in S}\overline{x}_i(\overline{v})$ and $\overline{v}(S)$ $(S\subset N)$ satisfy the constraint:

$$\sum_{i\in S} x_{Ri}(\overline{v}) < v_L(S),$$

then according to Eq. (2.8), the satisfactory crisp equivalent form of Eq. (2.12) can be constructed as follows:

$$\begin{cases} \sum_{i \in S} x_{Ri}(\overline{v}) < v_L(S) \quad (S \subset N) \\ \sum_{i \in N} x_{Ri}(\overline{v}) = v_R(N) \\ \sum_{i \in N} x_{Li}(\overline{v}) = v_L(N) \\ x_{Ri}(\overline{v}) \geq x_{Li}(\overline{v}) \quad (i = 1, 2, \ldots, n), \end{cases} \tag{2.15}$$

where $x_{Ri}(\overline{v})$ and $x_{Li}(\overline{v})$ $(i = 1, 2, \ldots, n)$ are decision variables.

Equation (2.15) is a system of linear inequalities, which can be solved by using the method of the system of inequalities.

For the situation in which the intervals $\sum_{i \in S} \overline{x}_i(\overline{v})$ and $\overline{v}(S)$ $(S \subset N)$ satisfy the constraint as follows:

$$\sum_{i \in S} x_{Li}(\overline{v}) < v_L(S) \leq \sum_{i \in S} x_{Ri}(\overline{v}) < v_R(S)$$

then according to Eq. (2.9), the satisfactory crisp equivalent form of Eq. (2.12) can be constructed as follows:

$$\begin{cases} \sum_{i \in S} x_{Li}(\overline{v}) < v_L(S) \quad (S \subset N) \\ \sum_{i \in S} x_{Ri}(\overline{v}) \geq v_L(S) \quad (S \subset N) \\ \sum_{i \in S} x_{Ri}(\overline{v}) < v_R(S) \quad (S \subset N) \\ \sum_{i \in N} x_{Ri}(\overline{v}) = v_R(N) \\ \sum_{i \in N} x_{Li}(\overline{v}) = v_L(N) \\ x_{Ri}(\overline{v}) \geq x_{Li}(\overline{v}) \quad (i = 1, 2, \ldots, n), \end{cases} \tag{2.16}$$

which is a system of linear inequalities about the decision variables $x_{Ri}(\overline{v})$ and $x_{Li}(\overline{v})$ $(i = 1, 2, \ldots, n)$.

For the situation in which the intervals $\sum_{i \in S} \overline{x}_i(\overline{v})$ and $\overline{v}(S)$ $(S \subset N)$ satisfy the constraints:

$$\sum_{i \in S} x_{Li}(\overline{v}) = v_L(S)$$

and

$$w\left(\sum_{i\in S}\overline{x}_i\right) = w(\overline{v}(S)),$$

then according to Eq. (2.10), the satisfactory crisp equivalent form of Eq. (2.12) can be constructed as follows:

$$\begin{cases} \sum_{i\in S} x_{Li}(\overline{v}) = v_L(S) \quad (S \subset N) \\ \sum_{i\in S} x_{Ri}(\overline{v}) - \sum_{i\in S} x_{Li}(\overline{v}) = v_R(S) - v_L(S) \quad (S \subset N) \\ \sum_{i\in N} x_{Ri}(\overline{v}) = v_R(N) \\ \sum_{i\in N} x_{Li}(\overline{v}) = v_L(N) \\ x_{Ri}(\overline{v}) \geq x_{Li}(\overline{v}) \quad (i = 1, 2, \ldots, n), \end{cases} \tag{2.17}$$

which is a system of linear inequalities about the decision variables $x_{Ri}(\overline{v})$ and $x_{Li}(\overline{v})$ $(i = 1, 2, \ldots, n)$.

2.3.3 Bisection Algorithm for Computing Interval-Valued Cores of Interval-Valued Cooperative Games

Solving Eq. (2.14) (or Eqs. (2.15)–(2.17)), we can obtain its optimal solution, denoted by $(\beta^*(\overline{v}), \overline{x}^*(\overline{v}))$ (or $\overline{x}^*(\overline{v})$). Then, $\overline{x}^*(\overline{v})$ is an element of the interval-valued core $\overline{C}(\overline{v})$ of the interval-valued cooperative game $\overline{v} \in \overline{G}^n$, where the maximum satisfactory degree is $\beta^*(\overline{v})$.

Obviously, if $\beta^*(\overline{v}) = 1$, then we can obtain the element of the interval-valued core $\overline{C}(\overline{v})$, which means that all the interval-valued inequalities in Eq. (2.14) are absolutely satisfied.

Generally, $(\beta^*(\overline{v}), \overline{x}^*(\overline{v}))$ is not a global optimal solution due to the fact that Eq. (2.14) is nonlinear programming. In the following, we propose the bisection method and algorithm for solving Eq. (2.14), which can always provide a global optimal solution to Eq. (2.14).

Assume that a precision $\varepsilon \in (0, 1]$ is given a priori. Then, according to the bisection method [28], we can determine the iteration number of the proposed bisection algorithm, denoted by m_0, where m_0 is a positive integer which is not smaller than $-\ln\varepsilon/\ln 2$. The bisection procedure and algorithm for solving Eq. (2.14) are summarized as follows:

Step 1: Let $t = 0$, and take $\beta_{Rt}(\overline{v}) = 1$. The nonlinear programming model (i.e., Eq. (2.14)) can be transformed into the linear programming. In this case, solving Eq. (2.14) with $\beta_{Rt}(\overline{v}) = 1$ by using the LINGO tool (or the simplex method of linear programming), if Eq. (2.14) has a feasible solution $\overline{\boldsymbol{x}}_t^*(\overline{v})$, then $\beta^*(\overline{v}) = \beta_{Rt}(\overline{v}) = 1$ is the optimal value of the objective function of Eq. (2.14) and $\overline{\boldsymbol{x}}^*(\overline{v}) = \overline{\boldsymbol{x}}_t^*(\overline{v})$ is an element of the interval-valued core $\overline{C}(\overline{v})$ with the maximum satisfactory degree $\beta^*(\overline{v})$. The algorithm stops. On the contrary, if there is not any feasible solution to Eq. (2.14), then go to Step 2.

Step 2: Take $\beta_{Lt}(\overline{v}) = 0$, and solve Eq. (2.14) by using the LINGO tool (or the simplex method of linear programming), if there is not any feasible solution to Eq. (2.14), which means that this linear programming (hereby Eq. (2.14)) has no optimal solutions, then the algorithm stops. On the contrary, if Eq. (2.14) with $\beta_{Lt}(\overline{v}) = 0$ has a feasible solution $\overline{\boldsymbol{x}}_t^*(\overline{v})$, then we can judge that the optimal value of the objective function of Eq. (2.14) is between 0 and 1, i.e., $\beta^*(v) \in (0, 1)$, go to Step 3.

Step 3: Let $m(\overline{\beta}_t(\overline{v}))$ be the mean of the lower bound $\beta_{Lt}(\overline{v})$ and the upper bound $\beta_{Rt}(\overline{v})$ of the interval $\overline{\beta}_t(\overline{v}) = [\beta_{Lt}(\overline{v}), \beta_{Rt}(\overline{v})]$. Namely,

$$m(\overline{\beta}_t(\overline{v})) = \frac{\beta_{Lt}(\overline{v}) + \beta_{Rt}(\overline{v})}{2} = \frac{0+1}{2} = 0.5.$$

By using the LINGO tool (or the simplex method of linear programming), solving Eq. (2.14) with $m(\overline{\beta}_t(\overline{v}))$, if there is not any feasible solution to Eq. (2.14), then the optimal value of the objective function of Eq. (2.14) falls into the range which is between the lower bound $\beta_{Lt}(\overline{v})$ and the mean $m(\overline{\beta}_t(\overline{v}))$ of the interval $\overline{\beta}_t(\overline{v})$, i.e., $\beta^*(\overline{v}) \in [\beta_{Lt}(\overline{v}), m(\overline{\beta}_t(\overline{v}))] = [0, 0.5]$, thereby the interval $\overline{\beta}_t(\overline{v})$ is narrowed. Let $\beta_{R,t+1}(\overline{v}) = m(\overline{\beta}_t(\overline{v})) = 0.5$ and $\beta_{L,t+1}(\overline{v}) = \beta_{Lt}(\overline{v}) = 0$, then go to Step 4. On the contrary, if Eq. (2.14) has a feasible solution $\overline{\boldsymbol{x}}_t^*(\overline{v})$, then the optimal value of the objective function of Eq. (2.14) falls into the range which is between the mean $m(\overline{\beta}_t(\overline{v}))$ and the upper bound $\beta_{Rt}(\overline{v})$ of the interval $\overline{\beta}_t(\overline{v})$, i.e., $\beta^*(\overline{v}) \in [m(\overline{\beta}_t(\overline{v})), \beta_{Rt}(\overline{v})] = [0.5, 1]$, thereby the interval $\overline{\beta}_t(\overline{v})$ is narrowed also. Let $\beta_{L,t+1}(\overline{v}) = m(\overline{\beta}_t(\overline{v})) = 0.5$ and $\beta_{R,t+1}(\overline{v}) = \beta_{Rt}(\overline{v}) = 1$, then go to Step 4.

Step 4: Let $t := t + 1$. Repeat Step 3 in the new smaller interval $\overline{\beta}_t(\overline{v}) = [\beta_{Lt}(\overline{v}), \beta_{Rt}(\overline{v})]$ until the m_0th iteration. Then, go to Step 5.

Step 5: The length of the narrowed interval $\overline{\beta}_{m_0}(\overline{v}) = [\beta_{Lm_0}(\overline{v}), \beta_{Rm_0}(\overline{v})]$ of the m_0th iteration is not greater than the given precision ε. Let

$$\beta^*(\overline{v}) \quad \frac{\beta_{Lm_0}(\overline{v}) + \beta_{Rm_0}(\overline{v})}{2}$$

which is the mean of the lower and upper bounds of the interval $\overline{\beta}_{m_0}(\overline{v})$. Accordingly, $\beta^*(\overline{v})$ is the optimal value of the objective function of Eq. (2.14) at a given precision ε; $\overline{\boldsymbol{x}}^*(\overline{v}) = \overline{\boldsymbol{x}}^*_{m_0}(\overline{v})$ is the element of the interval-valued core $\overline{C}(\overline{v})$ with the maximum satisfactory degree $\beta^*(\overline{v})$.

2.4 Real Example Analysis

Suppose that there are three companies p_1, p_2, and p_3 in the electronic product supply chain. They cooperate to develop a new type of electronic products. Each company has different superior resources. Due to a lack of information and/or imprecision of the available information, the managers of the three companies usually are not able to exactly forecast the profit amount of the companies' product under cooperation. Usually, the companies can predict the optimistic profit and the pessimistic profit of product. Hence, intervals are suitable to represent the profit amount of the product from the three companies' perspectives. If the three companies work together for product cooperative innovation, then the optimistic profit of the product may be 44 while the pessimistic profit of the product may be 40, which can be described as the interval-valued profit $\overline{v}'(1,2,3) = [40,44]$, where the numbers 1, 2, and 3 represent the companies p_1, p_2, and p_3 for short, respectively. Similarly, if the companies p_1 and p_2 cooperate for product innovation, then the interval-valued profit may be $\overline{v}'(1,2) = [22,30]$. If the companies p_1 and p_3 cooperate for product innovation, then the interval-valued profit may be $\overline{v}'(1,3) = [24,28]$. If the companies p_2 and p_3 cooperate for product innovation, then the interval-valued profit may be $\overline{v}'(2,3) = [20,32]$. Because of the limitation of resources, the three companies cannot develop and produce the product alone. Therefore, the profit of the product for each company is 0, i.e., $\overline{v}'(1) = \overline{v}'(2) = \overline{v}'(3) = 0$. Thus, the above problem may be regarded as an interval-valued cooperative game $\overline{v}' \in \overline{G}^3$. Namely, the three companies p_1, p_2, and p_3 in the electronic product supply chain may be regarded as the players 1, 2, and 3, respectively. The interval-valued characteristic function is $\overline{v}'$, which is defined on the grand coalition $N' = \{1,2,3\}$ so that $\overline{v}'(1,2,3) = [40,44]$, $\overline{v}'(1,2) = [22,30]$, $\overline{v}'(1,3) = [24,28]$, $\overline{v}'(2,3) = [20,32]$, and $\overline{v}'(1) = \overline{v}'(2) = \overline{v}'(3) = 0$.

2.4.1 *Computational Results Obtained by the Nonlinear Programming Method*

According to Eq. (2.14), the nonlinear programming model can be constructed as follows:

$$\max\{\beta(\overline{v}')\}$$

$$\text{s.t.}\begin{cases} x_{L1}(\overline{v}') + x_{L2}(\overline{v}') \geq 22 \\ x_{R1}(\overline{v}') + x_{R2}(\overline{v}') \leq 30 \\ x_{L1}(\overline{v}') + x_{L3}(\overline{v}') \geq 24 \\ x_{R1}(\overline{v}') + x_{R3}(\overline{v}') \leq 28 \\ x_{L2}(\overline{v}') + x_{L3}(\overline{v}') \geq 20 \\ x_{R2}(\overline{v}') + x_{R3}(\overline{v}') \leq 32 \\ (1-\beta(\overline{v}'))(x_{L1}(\overline{v}') + x_{L2}(\overline{v}')) + \beta(\overline{v}')(x_{R1}(\overline{v}') + x_{R2}(\overline{v}')) \geq 22(1-\beta(\overline{v}')) + 30\beta(\overline{v}') \\ (1-\beta(\overline{v}'))(x_{L1}(\overline{v}') + x_{L3}(\overline{v}')) + \beta(\overline{v}')(x_{R1}(\overline{v}') + x_{R3}(\overline{v}')) \geq 24(1-\beta(\overline{v}')) + 28\beta(\overline{v}') \\ (1-\beta(\overline{v}'))(x_{L2}(\overline{v}') + x_{L3}(\overline{v}')) + \beta(\overline{v}')(x_{R2}(\overline{v}') + x_{R3}(\overline{v}')) \geq 20(1-\beta(\overline{v}')) + 32\beta(\overline{v}') \\ x_{R1}(\overline{v}') + x_{R2}(\overline{v}') + x_{R3}(\overline{v}') = 44 \\ x_{L1}(\overline{v}') + x_{L2}(\overline{v}') + x_{L3}(\overline{v}') = 40 \\ x_{Ri}(\overline{v}') \geq x_{Li}(\overline{v}') \quad (i = 1,2,3) \\ 0 \leq \beta(\overline{v}') \leq 1, \end{cases} \tag{2.18}$$

where $x_{Ri}(\overline{v}')$, $x_{Li}(\overline{v}')$ $(i = 1,2,3)$, and $\beta(\overline{v}')$ are decision variables.

Solving Eq. (2.18) by the bisection algorithm given in Sect. 2.3.3, we can narrow the interval $[0,1]$ in which $\beta(\overline{v}')$ belongs to and infer that the optimal value $\beta^*(\overline{v}') \in [0.875, 0.87506]$. Thus, we can obtain the global optimal solution $(\beta^*(\overline{v}'), \overline{x}^*(\overline{v}'))$ of Eq. (2.18) at a given precision, where $\beta^*(\overline{v}') = 0.875$, $\overline{x}_1^*(\overline{v}') = [9.5, 13.5]$, $\overline{x}_2^*(\overline{v}') = [16, 16]$, and $\overline{x}_3^*(\overline{v}') = [14.5, 14.5]$. Therefore, for the interval-valued cooperative game $\overline{v}' \in \overline{G}^3$, $\overline{x}^*(\overline{v}')$ is an element of its interval-valued core $\overline{C}(\overline{v}')$ with the maximum satisfactory degree $\beta^*(\overline{v}') = 0.875$. In other words, if the maximum satisfactory degree of $\sum_{i\in S}\overline{x}_i(\overline{v}') \geq \overline{v}'(S)$ for the three companies p_1, p_2, and p_3 in the electronic product supply chain is not greater than 0.875, then the interval-valued core of the interval-valued cooperative game $\overline{v}' \in \overline{G}^3$ exists and hereby the three companies may choose product cooperative innovation.

Analogously, according to Eq. (2.15), the system of linear inequalities can be constructed as follows:

$$\begin{cases} x_{R1}\left(\overline{v}'\right)+x_{R2}\left(\overline{v}'\right)<22 \\ x_{R1}\left(\overline{v}'\right)+x_{R3}\left(\overline{v}'\right)<24 \\ x_{R2}\left(\overline{v}'\right)+x_{R3}\left(\overline{v}'\right)<20 \\ x_{R1}\left(\overline{v}'\right)+x_{R2}\left(\overline{v}'\right)+x_{R3}\left(\overline{v}'\right)=44 \\ x_{L1}\left(\overline{v}'\right)+x_{L2}\left(\overline{v}'\right)+x_{L3}\left(\overline{v}'\right)=40 \\ x_{Ri}\left(\overline{v}'\right)\geq x_{Li}\left(\overline{v}'\right) \quad (i=1,2,3), \end{cases} \tag{2.19}$$

where $x_{Ri}\left(\overline{v}'\right)$ and $x_{Li}\left(\overline{v}'\right)$ $(i=1,2,3)$ are decision variables.

Solving Eq. (2.19) by using the LINGO tool, we find that there is no feasible solution of Eq. (2.19) and hereby the three companies may have not any cooperative desire for this situation.

According to Eqs. (2.16) and (2.17), the systems of linear inequalities can be constructed as follows:

$$\begin{cases} x_{L1}\left(\overline{v}'\right)+x_{L2}\left(\overline{v}'\right)<22 \\ x_{L1}\left(\overline{v}'\right)+x_{L3}\left(\overline{v}'\right)<24 \\ x_{L2}\left(\overline{v}'\right)+x_{L3}\left(\overline{v}'\right)<20 \\ x_{R1}\left(\overline{v}'\right)+x_{R2}\left(\overline{v}'\right)\geq 22 \\ x_{R1}\left(\overline{v}'\right)+x_{R3}\left(\overline{v}'\right)\geq 24 \\ x_{R2}\left(\overline{v}'\right)+x_{R3}\left(\overline{v}'\right)\geq 20 \\ x_{R1}\left(\overline{v}'\right)+x_{R2}\left(\overline{v}'\right)<30 \\ x_{R1}\left(\overline{v}'\right)+x_{R3}\left(\overline{v}'\right)<28 \\ x_{R2}\left(\overline{v}'\right)+x_{R3}\left(\overline{v}'\right)<32 \\ x_{R1}\left(\overline{v}'\right)+x_{R2}\left(\overline{v}'\right)+x_{R3}\left(\overline{v}'\right)=44 \\ x_{L1}\left(\overline{v}'\right)+x_{L2}\left(\overline{v}'\right)+x_{L3}\left(\overline{v}'\right)=40 \\ x_{Ri}\left(\overline{v}'\right)\geq x_{Li}\left(\overline{v}'\right) \quad (i=1,2,3) \end{cases} \tag{2.20}$$

and

$$\begin{cases} x_{L1}(\overline{v}') + x_{L2}(\overline{v}') = 22 \\ x_{L1}(\overline{v}') + x_{L3}(\overline{v}') = 24 \\ x_{L2}(\overline{v}') + x_{L3}(\overline{v}') = 20 \\ x_{R1}(\overline{v}') + x_{R2}(\overline{v}') - (x_{L1}(\overline{v}') + x_{L2}(\overline{v}')) = 30 - 22 \\ x_{R1}(\overline{v}') + x_{R3}(\overline{v}') - (x_{L1}(\overline{v}') + x_{L3}(\overline{v}')) = 28 - 24 \\ x_{R2}(\overline{v}') + x_{R3}(\overline{v}') - (x_{L2}(\overline{v}') + x_{L3}(\overline{v}')) = 32 - 20 \\ x_{R1}(\overline{v}') + x_{R2}(\overline{v}') + x_{R3}(\overline{v}') = 44 \\ x_{L1}(\overline{v}') + x_{L2}(\overline{v}') + x_{L3}(\overline{v}') = 40 \\ x_{Ri}(\overline{v}') \geq x_{Li}(\overline{v}') \quad (i = 1,2,3), \end{cases} \tag{2.21}$$

respectively, where $x_{Ri}(\overline{v}')$ and $x_{Li}(\overline{v}')$ $(i = 1,2,3)$ are decision variables.

Solving Eqs. (2.20) and (2.21) by using the LINGO tool, respectively, we find that there are no feasible solutions of Eqs. (2.20) and (2.21) and hereby the three companies may have not any cooperative desire for these situations.

2.4.2 Computational Results Obtained by the Moore's Order Relation Between Intervals

According to Eq. (2.12), we construct the system of linear inequalities as follows:

$$\begin{cases} \bar{x}_1(\overline{v}') + \bar{x}_2(\overline{v}') \geq [22, 30] \\ \bar{x}_1(\overline{v}') + \bar{x}_3(\overline{v}') \geq [24, 28] \\ \bar{x}_2(\overline{v}') + \bar{x}_3(\overline{v}') \geq [20, 32] \\ \bar{x}_1(\overline{v}') + \bar{x}_2(\overline{v}') + \bar{x}_3(\overline{v}') = [40, 44] \\ x_{Ri}(\overline{v}') \geq x_{Li}(\overline{v}') \quad (i = 1,2,3), \end{cases} \tag{2.22}$$

where $x_{Ri}(\overline{v}')$ and $x_{Li}(\overline{v}')$ $(i = 1,2,3)$ are decision variables.

Using the Moore's order relation between intervals, i.e., Eq. (1.4), Eq. (2.22) can be rewritten as the following system of inequalities:

$$\begin{cases} x_{R1}(\overline{v}') + x_{R2}(\overline{v}') \geq 30 \\ x_{L1}(\overline{v}') + x_{L2}(\overline{v}') \geq 22 \\ x_{R1}(\overline{v}') + x_{R3}(\overline{v}') \geq 28 \\ x_{L1}(\overline{v}') + x_{L3}(\overline{v}') \geq 24 \\ x_{R2}(\overline{v}') + x_{R3}(\overline{v}') \geq 32 \\ x_{L2}(\overline{v}') + x_{L3}(\overline{v}') \geq 20 \\ x_{R1}(\overline{v}') + x_{R2}(\overline{v}') + x_{R3}(\overline{v}') = 44 \\ x_{L1}(\overline{v}') + x_{L2}(\overline{v}') + x_{L3}(\overline{v}') = 40 \\ x_{Ri}(\overline{v}') \geq x_{Li}(\overline{v}') \quad (i = 1, 2, 3). \end{cases} \tag{2.23}$$

Solving Eq. (2.23) by using the LINGO tool, we find that there is no feasible solution of Eq. (2.23) and hereby the three companies may have not any desire for product cooperative innovation.

Therefore, the interval-valued core of the interval-valued cooperative game $\overline{v}'$ $\in \overline{G}^3$ does not exist if the Moore's order relation between intervals is used. On the contrary, we can obtain an element of the interval-valued core of the interval-valued cooperative game $\overline{v}' \in \overline{G}^3$ by introducing the satisfactory degrees of comparing intervals. That is to say, the interval-valued core $\overline{C}(\overline{v}')$ of the interval-valued cooperative game $\overline{v}' \in \overline{G}^3$ exists, i.e., $\overline{C}(\overline{v}') \neq \varnothing$. This result may give more management suggestions for the players (or managers).

References

1. Li D-F. Fuzzy multiobjective many-person decision makings and games. Beijing: National Defense Industry Press; 2003 (in Chinese).
2. Owen G. Game theory. 2nd ed. New York: Academic Press; 1982.
3. Branzei R, Branzei O, Alparslan Gök SZ, Tijs S. Cooperative interval games: a survey. Cent Eur J Oper Res. 2010;18:397–411.
4. Gillies DB. Solutions to general non-zero-sum games. In: Tucker AW, Luce RD, editors. Contributions to theory of games IV, Annals of mathematical studies, vol. 40. Princeton: Princeton University Press; 1959. p. 47–85.
5. Driessen T. Cooperation games: solutions and application. Netherlands: Kluwer Academic Publisher; 1988.
6. Shapley L. Cores of convex games. Int J Game Theory. 1971;1:11–26.
7. Branzei R, Alparslan-Gök SZ, Branzei O. Cooperation games under interval uncertainty: on the convexity of the interval undominated cores. Cent Eur J Oper Res. 2011;19:523–32.
8. Alparslan-Gök SZ, Branzei R, Tijs SH. Cores and stable sets for interval-valued games, vol. 1. Center for Economic Research, Tilburg University; 2008. p. 1–14.

9. Alparslan-Gök SZ, Branzei O, Branzei R, Tijs S. Set-valued solution concepts using interval-type payoffs for interval games. J Math Econ. 2011;47:621–6.
10. Shapley LS. On balanced sets and cores. Naval Res Logist Quart. 1967;14:453–60.
11. Han W-B, Sun H, Xu G-J. A new approach of cooperative interval games: the interval core and Shapley value revisited. Oper Res Lett. 2012;40:462–8.
12. Moore R. Methods and applications of interval analysis. Philadelphia: SIAM Studies in Applied Mathematics; 1979.
13. Li D-F. Linear programming approach to solve interval-valued matrix games. Omega. 2011;39 (6):655–66.
14. Sengupta A, Pal TK. Theory and methodology on comparing interval numbers. Eur J Oper Res. 2000;127:28–43.
15. Ishihuchi H, Tanaka M. Multiobjective programming in optimization of the interval objective function. Eur J Oper Res. 1990;48:219–25.
16. Li D-F, Nan J-X, Zhang M-J. Interval programming models for matrix games with interval payoffs. Optim Methods Softw. 2012;27:1–16.
17. Zadeh L. Fuzzy sets. Inform Control. 1965;8:338–56.
18. Dubois D, Prade H. Fuzzy sets and systems: theory and applications. New York: Academic Press; 1980.
19. Collins WD, Hu C-Y. Interval matrix games. In: Hu C-Y, Kearfott RB, Korvinet AD, et al., editors. Knowledge processing with interval and soft computing. London: Springer; 2008. p. 168–72.
20. Collins WD, Hu C-Y. Studying interval valued matrix games with fuzzy logic. Soft Comput. 2008;12(2):147–55.
21. Nayak PK, Pal M. Linear programming technique to solve two person matrix games with interval pay-offs. Asia Pac J Oper Res. 2009;26(2):285–305.
22. Li D-F. Notes on "linear programming technique to solve two person matrix games with interval pay-offs". Asia Pac J Oper Res. 2011;28(6):705–37.
23. Zimmermann H-J. Fuzzy set theory and its application. 2nd ed. Dordrecht: Kluwer Academic Publishers; 1991.
24. Li D-F. Lexicographic method for matrix games with payoffs of triangular fuzzy numbers. Int J Uncertain Fuzziness Knowl Based Syst. 2008;16(3):371–89.
25. Gillies DB. Some theorems on n-person games. PhD thesis. Princeton: Princeton University Press; 1953.
26. Branzei R, Dimitrov D, Tijs S. Models in cooperative game theory. Game theory and mathematical methods. Berlin: Springer; 2008.
27. Alparslan-Gök SZ, Branzei R, Tijs S. Big boss interval games. Institute of Applied Mathematics, METU and Tilburg University, Center for Economic Research, The Netherlands, CentER DP 47 (preprint no. 103); 2008.
28. Sikorski K. Bisection is optimal. Numer Math. 1982;40:111–7.

Chapter 3
Several Interval-Valued Solutions of Interval-Valued Cooperative Games and Simplified Methods

Abstract The aim of this chapter is to develop direct and effective simplified methods for computing interval-valued cooperative games. In this chapter, we propose several commonly used and important concepts of interval-valued solutions such as the interval-valued equal division value, the interval-valued equal surplus division value, the interval-valued Shapley value, the interval-valued egalitarian Shapley value, the interval-valued discounted Shapley value, the interval-valued solidarity value, and the interval-valued generalized solidarity value as well as the interval-valued Banzhaf value. Through adding some conditions such as the size monotonicity, we prove that the aforementioned corresponding solutions of cooperative games are continuous, monotonic, and non-decreasing functions of coalitions' values. Hereby, the aforementioned interval-valued solutions of interval-valued cooperative games can be directly and explicitly obtained by determining their lower and upper bounds, respectively. Moreover, we discuss these interval-valued solutions' important properties. Thus, we may overcome the issues of the Moore's interval subtraction. The feasibility and applicability of the methods proposed in this chapter are illustrated with real numerical examples.

Keywords Interval-valued cooperative game • Interval-valued equal surplus division value • Interval-valued Shapley value • Interval-valued solidarity value • Interval-valued Banzhaf value

3.1 Introduction

In the preceding Chaps. 1 and 2, we have formulated the interval-valued cooperative games [1, 2], which are a natural extension of cooperative games [3, 4]. Two important concepts, i.e., the interval-valued least square solution and the satisfactory interval-valued core, are proposed for interval-valued cooperative games and corresponding quadratic programming models and auxiliary nonlinear programming models are established, respectively. The aforementioned interval-valued solutions and the derived mathematical programming models do not use the Moore's interval subtraction [5]. Thus, they can overcome the irrational issues [6, 7] resulted from the interval subtraction operator. For example, the Moore's interval subtraction is not invertible. Indeed, the interval subtraction and interval

D.-F. Li, *Models and Methods for Interval-Valued Cooperative Games in Economic Management*, DOI 10.1007/978-3-319-28998-4_3

ranking (or comparison) are two important problems in interval-valued cooperative game theory and application [8, 9]. Therefore, in this chapter, we still investigate on how to solve interval-valued cooperative games without using either the interval subtraction or the ranking of intervals. More specifically, for a kind of interval-valued cooperative games with the size monotonicity, we will define several commonly used and important concepts of interval-valued solutions such as the interval-valued equal division value, the interval-valued equal surplus division value, the interval-valued Shapley value, the interval-valued egalitarian Shapley value, the interval-valued discounted Shapley value, the interval-valued solidarity value, and the interval-valued generalized solidarity value as well as the interval-valued Banzhaf value. Thus, through adding some conditions such as the size monotonicity, it is proven that the aforementioned corresponding solutions of cooperative games are continuous, monotonic, and non-decreasing. Hereby these interval-valued solutions of interval-valued cooperative games can be directly and explicitly obtained by determining their lower and upper bounds, respectively. Moreover, it is proven that the derived interval-valued solutions possess some useful and important properties.

The rest of this chapter is organized as follows. Section 3.2 studies the interval-valued equal division value and the interval-valued equal surplus division value of interval-valued cooperative games and their properties. Section 3.3 gives the interval-valued Shapley value, the interval-valued egalitarian Shapley value, and the interval-valued discounted Shapley value of interval-valued cooperative games and their properties. In Sect. 3.4, we discuss the interval-valued solidarity value and the interval-valued generalized solidarity value of interval-valued cooperative games and their properties. Section 3.5 investigates the interval-valued Banzhaf value of interval-valued cooperative games and its properties.

3.2 Interval-Valued Equal Division Values and Interval-Valued Equal Surplus Division Values of Interval-Valued Cooperative Games

For an arbitrary cooperative game $\upsilon \in G^n$ stated as in the previous Sect. 1.2, we can define its equal division value (or solution) as

$$\boldsymbol{\rho}^{\mathrm{ED}}(\upsilon) = \left(\rho_1^{\mathrm{ED}}(\upsilon), \rho_2^{\mathrm{ED}}(\upsilon), \ldots, \rho_n^{\mathrm{ED}}(\upsilon)\right)^{\mathrm{T}},$$

whose components are given as follows:

$$\rho_i^{\mathrm{ED}}(\upsilon) = \frac{\upsilon(N)}{n} \quad (i = 1, 2, \ldots, n), \tag{3.1}$$

which means that the worth $\upsilon(N)$ of the grand coalition N is distributed equally among all n players in the cooperative game $\upsilon \in G^n$.

For reasons that are clear from Eq. (3.1), the equal division value sometime is called the egalitarian value (or solution).

It is easily proven that the equal division value $\boldsymbol{\rho}^{\text{ED}}(v)$ of any cooperative game $v \in G^n$ satisfies the efficiency, the symmetry, and the additivity [9, 10]. These properties also are referred to those similarly stated as in the previous Sect. 1.4.1.

There is a related solution, i.e., the equal surplus division value (or solution), which can be defined as

$$\boldsymbol{\rho}^{\text{ESD}}(v) = \left(\rho_1^{\text{ESD}}(v), \rho_2^{\text{ESD}}(v), \ldots, \rho_n^{\text{ESD}}(v)\right)^{\text{T}},$$

whose components are given as follows:

$$\rho_i^{\text{ESD}}(v) = v(i) + \frac{v(N) - \sum_{j=1}^{n} v(j)}{n} \quad (i = 1, 2, \ldots, n), \tag{3.2}$$

which means that the individual worth $v(i)$ is firstly assigned to the player $i \in N$ and then the remainder of $v(N)$, i.e., $v(N) - \sum_{j=1}^{n} v(j)$, is distributed equally among all n players in the cooperative game $v \in G^n$. Sometimes, the equal surplus division value is also called as the center of the imputation set (or shortly CIS) value [11, 12].

In the same way, it is easily proven that the equal surplus division value $\boldsymbol{\rho}^{\text{ESD}}(v)$ of any cooperative game $v \in G^n$ satisfies the efficiency, the symmetry, and the additivity [9, 10]. Furthermore, if a cooperative game $v \in G^n$ satisfies the following condition:

$$v(N) \geq \sum_{i=1}^{n} v(i),$$

then its equal surplus division value $\boldsymbol{\rho}^{\text{ESD}}(v)$ satisfies the individual rationality, i.e.,

$$\rho_i^{\text{ESD}}(v) \geq v(i) \quad (i = 1, 2, \ldots, n).$$

Particularly, for any superadditive cooperative game $v \in G^n$, it is easy to prove that its equal surplus division value is an imputation of the cooperative game v, i.e.,

$$\boldsymbol{\rho}^{\text{ESD}}(v) \in I(v).$$

3.2.1 Interval-Valued Equal Division Values of Interval-Valued Cooperative Games and Simplified Methods

Usually, for any interval-valued cooperative game $\bar{v} \in \overline{G}^n$ stated as in the previous Sect. 1.3.2, the payoff of each player $i \in N$ should be an interval also. Consequently, in a similar way to Eq. (3.1), we can easily define an interval-valued equal division value $\overline{\boldsymbol{\rho}}^{\mathrm{ED}}(\overline{v}) = \left(\overline{\rho}_1^{\mathrm{ED}}(\overline{v}), \overline{\rho}_2^{\mathrm{ED}}(\overline{v}), \ldots, \overline{\rho}_n^{\mathrm{ED}}(\overline{v})\right)^{\mathrm{T}}$ of any interval-valued cooperative game $\overline{v} \in \overline{G}^n$ according to the case 3 of Definition 1.1, where

$$\overline{\rho}_i^{\mathrm{ED}}(\overline{v}) = \frac{\overline{v}(N)}{n} \quad (i = 1, 2, \ldots, n).$$

Namely,

$$\overline{\rho}_i^{\mathrm{ED}}(\overline{v}) = \left[\frac{v_L(N)}{n}, \frac{v_R(N)}{n}\right] \quad (i = 1, 2, \ldots, n). \tag{3.3}$$

Alternatively, Eq. (3.3) may be obtained in the following way. For the interval-valued cooperative game $\overline{v} \in \overline{G}^n$, we can define an associated cooperative game $v(\alpha) \in G^n$, where the set of players still is $N = \{1, 2, \ldots, n\}$ and the characteristic function $v(\alpha)$ of coalitions of players is defined as follows:

$$v(\alpha)(S) = (1 - \alpha)v_L(S) + \alpha v_R(S) \quad (S \subseteq N) \tag{3.4}$$

and $v(\alpha)(\varnothing) = 0$. The parameter $\alpha \in [01]$ is any real number, which may be interpreted as an attitude factor [13].

According to Eq. (3.1), we can easily obtain the equal division value $\boldsymbol{\rho}^{\mathrm{ED}}(v(\alpha)) = \left(\rho_1^{\mathrm{ED}}(v(\alpha)), \rho_2^{\mathrm{ED}}(v(\alpha)), \ldots, \rho_n^{\mathrm{ED}}(v(\alpha))\right)^{\mathrm{T}}$ of the cooperative game $v(\alpha) \in G^n$, where

$$\rho_i^{\mathrm{ED}}(v(\alpha)) = \frac{v(\alpha)(N)}{n} \quad (i = 1, 2, \ldots, n),$$

i.e.,

$$\rho_i^{\mathrm{ED}}(v(\alpha)) = \frac{(1 - \alpha)v_L(N) + \alpha v_R(N)}{n} \quad (i = 1, 2, \ldots, n). \tag{3.5}$$

Obviously, the equal division value $\boldsymbol{\rho}^{\mathrm{ED}}(v(\alpha))$ is a continuous function of $v(\alpha)$. Note that $v(\alpha)$ is also a continuous function of $\alpha \in [0, 1]$ due to Eq. (3.4). Therefore, the equal division value $\boldsymbol{\rho}^{\mathrm{ED}}(v(\alpha))$ is a continuous function of the parameter $\alpha \in [0, 1]$.

Furthermore, for any $\alpha \in [0, 1]$ and $\alpha' \in [0, 1]$, if $\alpha \geq \alpha'$, i.e., $v(\alpha)(S) \geq v(\alpha')(S)$ for any coalition $S \subseteq N$, then particularly we have

$$v(\alpha)(N) \geq v\left(\alpha'\right)(N).$$

Hereby, it easily follows from Eq. (3.5) that

$$\rho_i^{\mathrm{ED}}(v(\alpha)) \geq \rho_i^{\mathrm{ED}}\left(v\left(\alpha'\right)\right) \quad (i = 1, 2, \ldots, n).$$

Therefore, the equal division value $\rho^{\mathrm{ED}}(v(\alpha))$ is a monotonic and non-decreasing function of the parameter $\alpha \in [0,1]$.

Thus, the lower and upper bounds of the interval-valued equal division value $\overline{\boldsymbol{\rho}}^{\mathrm{ED}}(\overline{v}) = (\overline{\rho}_1^{\mathrm{ED}}(\overline{v}), \overline{\rho}_2^{\mathrm{ED}}(\overline{v}), \ldots, \overline{\rho}_n^{\mathrm{ED}}(\overline{v}))^{\mathrm{T}}$ of the interval-valued cooperative game $\overline{v} \in \overline{G}^n$ can be attained at the lower and upper bounds of the interval $[0,1]$, respectively, i.e.,

$$\rho_{Li}^{\mathrm{ED}}(\overline{v}) = \rho_i^{\mathrm{ED}}(v(0)) = \frac{v_L(N)}{n} \quad (i = 1, 2, \ldots, n)$$

and

$$\rho_{Ri}^{\mathrm{ED}}(\overline{v}) = \rho_i^{\mathrm{ED}}(v(1)) = \frac{v_R(N)}{n} \quad (i = 1, 2, \ldots, n).$$

Thus, we have

$$\overline{\rho}_i^{\mathrm{ED}}(\overline{v}) = \left[\rho_{Li}^{\mathrm{ED}}(\overline{v}), \rho_{Ri}^{\mathrm{ED}}(\overline{v})\right] = \left[\frac{v_L(N)}{n}, \frac{v_R(N)}{n}\right] \quad (i = 1, 2, \ldots, n),$$

which are just about Eq. (3.3).

Example 3.1 Let us compute the interval-valued equal division value of the interval-valued cooperative game $\overline{v}' \in \overline{G}^3$ given in Example 1.1, where $N' = \{1,2,3\}$.

From Example 1.1, we have $\overline{v}'(1,2,3) = [6,7]$. According to Eq. (3.3), we can easily obtain the interval-valued equal division value of the interval-valued cooperative game $\overline{v}' \in \overline{G}^3$ as follows:

$$\begin{aligned}
\overline{\boldsymbol{\rho}}^{\mathrm{ED}}(\overline{v}') &= \left(\overline{\rho}_1^{\mathrm{ED}}(\overline{v}'), \overline{\rho}_2^{\mathrm{ED}}(\overline{v}'), \overline{\rho}_3^{\mathrm{ED}}(\overline{v}')\right)^{\mathrm{T}} \\
&= \left(\frac{\overline{v}'(1,2,3)}{3}, \frac{\overline{v}'(1,2,3)}{3}, \frac{\overline{v}'(1,2,3)}{3}\right)^{\mathrm{T}} \\
&= \left(\left[2, \frac{7}{3}\right], \left[2, \frac{7}{3}\right], \left[2, \frac{7}{3}\right]\right)^{\mathrm{T}}.
\end{aligned}$$

Apparently, each player can get the identical interval-valued payoff [2, 7/3].

In the following, we discuss some useful and important properties of interval-valued equal division values of interval-valued cooperative games.

Theorem 3.1 *(Existence and Uniqueness) For an arbitrary interval-valued cooperative game $\overline{v} \in \overline{G}^n$, there always exists a unique interval-valued equal division value $\overline{\boldsymbol{\rho}}^{\mathrm{ED}}(\overline{v})$, which is determined by Eq. (3.3).*

Proof According to Eq. (3.3), it is easy to prove that Theorem 3.1 is valid.

Theorem 3.2 *(Efficiency) For any interval-valued cooperative game $\overline{v} \in \overline{G}^n$, then its interval-valued equal division value $\overline{\boldsymbol{\rho}}^{\mathrm{ED}}(\overline{v})$ satisfies the efficiency, i.e., $\sum_{i=1}^{n} \overline{\rho}_i^{\mathrm{ED}}(\overline{v}) = \overline{v}(N)$.*

Proof According to Eq. (3.3), and combining with Definition 1.1, we have

$$\begin{aligned}\sum_{i=1}^{n} \overline{\rho}_i^{\mathrm{ED}}(\overline{v}) &= \sum_{i=1}^{n} \left[\frac{v_L(N)}{n}, \frac{v_R(N)}{n}\right] \\ &= [v_L(N), v_R(N)] \\ &= \overline{v}(N),\end{aligned}$$

i.e.,

$$\sum_{i=1}^{n} \overline{\rho}_i^{\mathrm{ED}}(\overline{v}) = \overline{v}(N).$$

Thus, we have completed the proof of Theorem 3.2.

Theorem 3.3 *(Additivity) For any two interval-valued cooperative games $\overline{v} \in \overline{G}^n$ and $\overline{\nu} \in \overline{G}^n$, then $\overline{\rho}_i^{\mathrm{ED}}(\overline{v} + \overline{\nu}) = \overline{\rho}_i^{\mathrm{ED}}(\overline{v}) + \overline{\rho}_i^{\mathrm{ED}}(\overline{\nu})$ $(i = 1, 2, \ldots, n)$, i.e., $\overline{\boldsymbol{\rho}}^{\mathrm{ED}}(\overline{v} + \overline{\nu}) = \overline{\boldsymbol{\rho}}^{\mathrm{ED}}(\overline{v}) + \overline{\boldsymbol{\rho}}^{\mathrm{ED}}(\overline{\nu})$.*

Proof According to Eq. (3.3) and Definition 1.1, we have

$$\begin{aligned}\overline{\rho}_i^{\mathrm{ED}}(\overline{v} + \overline{\nu}) &= \frac{\overline{v}(N) + \overline{\nu}(N)}{n} \\ &= \frac{\overline{v}(N)}{n} + \frac{\overline{\nu}(N)}{n} \\ &= \overline{\rho}_i^{\mathrm{ED}}(\overline{v}) + \overline{\rho}_i^{\mathrm{ED}}(\overline{\nu}),\end{aligned}$$

i.e.,

$$\overline{\rho}_i^{\mathrm{ED}}(\overline{v} + \overline{\nu}) = \overline{\rho}_i^{\mathrm{ED}}(\overline{v}) + \overline{\rho}_i^{\mathrm{ED}}(\overline{\nu}) \quad (i = 1, 2, \ldots, n).$$

Hence, we obtain

$$\overline{\boldsymbol{\rho}}^{\mathrm{ED}}(\overline{v} + \overline{\nu}) = \overline{\boldsymbol{\rho}}^{\mathrm{ED}}(\overline{v}) + \overline{\boldsymbol{\rho}}^{\mathrm{ED}}(\overline{\nu}).$$

Therefore, we have completed the proof of Theorem 3.3.

Theorem 3.4 *(Symmetry) For any interval-valued cooperative game* $\bar{v} \in \overline{G}^n$, *if* $i \in N$ *and* $k \in N$ $(i \neq k)$ *are two symmetric players in the interval-valued cooperative game* $\bar{v}$, *then* $\bar{\rho}_i^{\mathrm{ED}}(\bar{v}) = \bar{\rho}_k^{\mathrm{ED}}(\bar{v})$.

Proof It easily follows from Eq. (3.3) and Definition 1.3 that the conclusion of Theorem 3.4 is valid(omitted).

Theorem 3.5 *(Anonymity) For an arbitrary interval-valued cooperative game* $\bar{v} \in \overline{G}^n$ *and any permutation* σ *on the set* N, *then* $\bar{\rho}_{\sigma(i)}^{\mathrm{ED}}(\bar{v}^\sigma) = \bar{\rho}_i^{\mathrm{ED}}(\bar{v})$ $(i = 1, 2, \ldots, n)$. *Namely,* $\bar{\boldsymbol{\rho}}^{\mathrm{ED}}(\bar{v}^\sigma) = \boldsymbol{\sigma}^{\#}(\bar{\boldsymbol{\rho}}^{\mathrm{ED}}(\bar{v}))$.

Proof According to Eq. (3.3), we can easily prove Theorem 3.5 (omitted).

However, it is obvious from Eq. (3.3) that interval-valued equal division values of interval-valued cooperative games do not always satisfy the dummy player property and the null player property.

Example 3.2 Let us consider the simple interval-valued cooperative game $\bar{v}^0 \in \overline{G}^3$ as follows: $\bar{v}^0(N^0) = \bar{v}^0(2,3) = [3,6]$ and $\bar{v}^0(S) = 0$ for all other coalitions $S \subseteq N^0 = \{1,2,3\}$.

Note that $\bar{v}^0(1) = 0$. Therefore, it is obvious that

$$\bar{v}^0(S \cup 1) = \bar{v}^0(S)$$

and

$$\bar{v}^0(S \cup 1) = \bar{v}^0(S) + \bar{v}^0(1)$$

for any coalition $S \subseteq N^0 \backslash \{1\} = \{2,3\}$. Thus, the player 1 is not only a null player but also a dummy player in the above interval-valued cooperative game $\bar{v}^0 \in \overline{G}^3$. Intuitively, it seems to be reasonable that the player 1 in the above interval-valued cooperative game $\bar{v}^0 \in \overline{G}^3$ should obtain zero. However, according to Eq. (3.3), we can obtain the interval-valued equal division value of the interval-valued cooperative game $\bar{v}^0 \in \overline{G}^3$ as follows:

$$\begin{aligned}
\bar{\boldsymbol{\rho}}^{\mathrm{ED}}(\bar{v}^0) &= \left(\bar{\rho}_1^{\mathrm{ED}}(\bar{v}^0), \bar{\rho}_2^{\mathrm{ED}}(\bar{v}^0), \bar{\rho}_3^{\mathrm{ED}}(\bar{v}^0)\right)^{\mathrm{T}} \\
&= \left(\frac{\bar{v}^0(1,2,3)}{3}, \frac{\bar{v}^0(1,2,3)}{3}, \frac{\bar{v}^0(1,2,3)}{3}\right)^{\mathrm{T}} \\
&= ([1,2],[1,2],[1,2])^{\mathrm{T}}.
\end{aligned}$$

Accordingly, the player 1 can obtain the interval-valued payoff $[1,2]$ from the interval-valued cooperative game $\bar{v}^0 \in \overline{G}^3$. That is to say, the interval-valued equal division value $\bar{\boldsymbol{\rho}}^{\mathrm{ED}}(\bar{v}^0)$ does not satisfy the dummy player property and the null player property.

3.2.2 *Interval-Valued Equal Surplus Division Values of Interval-Valued Cooperative Games and Simplified Methods*

Theoretically, in a similar way to Eq. (3.3), we can employ Eq. (3.2) to define an interval-valued equal surplus division value $\overline{\rho}^{\text{ESD}}(\overline{v}) = (\overline{\rho}_1^{\text{ESD}}(\overline{v}), \overline{\rho}_2^{\text{ESD}}(\overline{v}), \ldots, \overline{\rho}_n^{\text{ESD}}(\overline{v}))^{\text{T}}$ of any interval-valued cooperative game $\overline{v} \in \overline{G}^n$. In this case, however, we have to use the interval subtraction operation such as the Moore's interval subtraction [5] or the Hukuhara difference [14]. As stated earlier, the interval subtraction may result in some irrational conclusions in that it is not an invertible operator [15]. Therefore, in what follows, we focus on developing a direct and an effective simplified method for computing interval-valued equal surplus division values of interval-valued cooperative games through using the monotonicity rather than the interval subtraction.

For any interval-valued cooperative game $\overline{v} \in \overline{G}^n$ stated as in Sect. 1.3.2, we can similarly construct an associated cooperative game $v(\alpha) \in G^n$, where the set of players is $N = \{1, 2, \ldots, n\}$ and the characteristic function $v(\alpha)$ of coalitions of players is given by Eq. (3.4).

According to Eq. (3.2), we can easily obtain the equal surplus division value $\boldsymbol{\rho}^{\text{ESD}}(v(\alpha)) = (\rho_1^{\text{ESD}}(v(\alpha)), \rho_2^{\text{ESD}}(v(\alpha)), \ldots, \rho_n^{\text{ESD}}(v(\alpha)))^{\text{T}}$ of the cooperative game $v(\alpha) \in G^n$, where

$$\rho_i^{\text{ESD}}(v(\alpha)) = v(\alpha)(i) + \frac{v(\alpha)(N) - \sum_{j=1}^{n} v(\alpha)(j)}{n} \quad (i = 1, 2, \ldots, n),$$

i.e.,

$$\rho_i^{\text{ESD}}(v(\alpha)) = (1-\alpha)v_L(i) + \alpha v_R(i) + \frac{(1-\alpha)v_L(N) + \alpha v_R(N) - \sum_{j=1}^{n}[(1-\alpha)v_L(j) + \alpha v_R(j)]}{n} \quad (i = 1, 2, \ldots, n). \tag{3.6}$$

It is obvious from Eq. (3.6) that the equal surplus division value $\boldsymbol{\rho}^{\text{ESD}}(v(\alpha))$ of the cooperative game $v(\alpha) \in G^n$ is a continuous function of the parameter $\alpha \in [0, 1]$.

Theorem 3.6 *For any interval-valued cooperative game $\overline{v} \in \overline{G}^n$, if the following system of inequalities*

$$v_R(N) - v_L(N) \geq \sum_{j=1}^{n} [(v_R(j) - v_L(j)) - (v_R(i) - v_L(i))] \quad (i = 1, 2, \ldots, n) \tag{3.7}$$

is satisfied, then the equal surplus division value $\boldsymbol{\rho}^{\text{ESD}}(v(\alpha))$ of the cooperative game $v(\alpha) \in G^n$ is a monotonic and non-decreasing function of the parameter $\alpha \in [0, 1]$.

Proof For any $\alpha \in [0,1]$ and $\alpha' \in [0,1]$, if $\alpha \geq \alpha'$, according to Eq. (3.6), and combining with the assumption, i.e., Eq. (3.7), we have

$$\begin{aligned}
&\rho_i^{\mathrm{ESD}}(v(\alpha)) - \rho_i^{\mathrm{ESD}}(v(\alpha')) \\
&= (\alpha - \alpha')\left[(v_R(i) - v_L(i)) + \frac{(v_R(N) - v_L(N)) - \sum_{j=1}^{n}(v_R(j) - v_L(j))}{n}\right] \\
&= (\alpha - \alpha')\frac{(v_R(N) - v_L(N)) - \sum_{j=1}^{n}[(v_R(j) - v_L(j)) - (v_R(i) - v_L(i))]}{n} \\
&\geq 0,
\end{aligned}$$

where $i = 1, 2, \ldots, n$. Hence, we have

$$\rho_i^{\mathrm{ESD}}(v(\alpha)) \geq \rho_i^{\mathrm{ESD}}\left(v\left(\alpha'\right)\right) \quad (i = 1, 2, \ldots, n),$$

which mean that the equal surplus division value $\rho^{\mathrm{ESD}}(v(\alpha))$ is a monotonic and non-decreasing function of the parameter $\alpha \in [0,1]$. We have completed the proof of Theorem 3.6.

Obviously, Eq. (3.7) can be written as the following system of inequalities:

$$\begin{aligned}
v_R(N) - v_L(N) \geq -n(v_R(i) - v_L(i)) \\
+\sum_{j=1}^{n}(v_R(j) - v_L(j)) \quad (i = 1, 2, \ldots, n).
\end{aligned} \tag{3.8}$$

Note that $v_R(i) \geq v_L(i)$ $(i = 1, 2, \ldots, n)$. Thus, the condition given by Eq. (3.8) is weaker than that expressed with the following system of inequalities:

$$v_R(N) - v_L(N) \geq \sum_{i=1}^{n}(v_R(i) - v_L(i)). \tag{3.9}$$

That is to say, if Eq. (3.9) is satisfied, then Eq. (3.8) is always true. Equations (3.8) and (3.9) are important conditions which are proposed in this section for the sequent study.

It is worthwhile to point out that even if both

$$v_R(N) \geq \sum_{i=1}^{n} v_R(i) \tag{3.10}$$

and

$$v_L(N) \geq \sum_{i=1}^{n} v_L(i), \tag{3.11}$$

i.e.,

$$\overline{v}(N) \geq \sum_{i=1}^{n} \overline{v}(i)$$

in the Moore's order relation or ranking method for intervals [5], Eq. (3.9) is not always true. Conversely, even if Eq. (3.9) is satisfied, Eqs. (3.10) and (3.11) are not always true at the same time.

If the interval-valued cooperative game $\overline{v} \in \overline{G}^n$ degenerates to a (classical) cooperative game $v \in G^n$, i.e., $v_R(S) = v_L(S)$ for all coalitions $S \subseteq N$, then Eq. (3.9) is reduced to the following system of inequalities:

$$v(N) \geq \sum_{i=1}^{n} v(i),$$

which can be derived from the superadditivity stated as in the previous Sect. 1.2, where $v(S) = v_R(S) = v_L(S)$. Thus, Eq. (3.7) may be similarly interpreted as follows: the length of the value (i.e., interval) of the grand coalition N is not smaller than the sum of the lengths of the values (i.e., intervals) of the individual players $i \in N$.

Therefore, for an interval-valued cooperative game $\overline{v} \in \overline{G}^n$, if it satisfies Eq. (3.7), then according to Theorem 3.6, the lower and upper bounds of the interval-valued equal surplus division value $\overline{\rho}^{\mathrm{ESD}}(\overline{v}) = (\overline{\rho}_1^{\mathrm{ESD}}(\overline{v}), \overline{\rho}_2^{\mathrm{ESD}}(\overline{v}), \ldots, \overline{\rho}_n^{\mathrm{ESD}}(\overline{v}))^{\mathrm{T}}$ can be attained at the lower and upper bounds of the interval [0, 1], respectively. Thus, according to Eq. (3.6), we have

$$\rho_{Li}^{\mathrm{ESD}}(\overline{v}) = \rho_i^{\mathrm{ESD}}(v(0)) = v_L(i) + \frac{v_L(N) - \sum_{j=1}^{n} v_L(j)}{n} \quad (i = 1, 2, \ldots, n)$$

and

$$\rho_{Ri}^{\mathrm{ESD}}(\overline{v}) = \rho_i^{\mathrm{ESD}}(v(1)) = v_R(i) + \frac{v_R(N) - \sum_{j=1}^{n} v_R(j)}{n} \quad (i = 1, 2, \ldots, n).$$

Namely,

$$\rho_{Li}^{\mathrm{ESD}}(\overline{v}) = v_L(i) + \frac{v_L(N) - \sum_{j=1}^{n} v_L(j)}{n} \quad (i = 1, 2, \ldots, n) \tag{3.12}$$

and

$$\rho_{Ri}^{\mathrm{ESD}}(\overline{v}) = v_R(i) + \frac{v_R(N) - \sum_{j=1}^{n} v_R(j)}{n} \quad (i = 1, 2, \ldots, n). \tag{3.13}$$

Then, we have

$$\overline{\rho}_i^{\mathrm{ESD}}(\overline{v}) = \left[v_L(i) + \frac{v_L(N) - \sum_{j=1}^{n} v_L(j)}{n}, v_R(i) + \frac{v_R(N) - \sum_{j=1}^{n} v_R(j)}{n} \right] \quad (i = 1, 2, \ldots, n), \tag{3.14}$$

which can be used to compute the interval-valued equal surplus division value $\overline{\rho}^{\mathrm{ESD}}(\overline{v})$ of the interval-valued cooperative game $\overline{v} \in \overline{G}^n$ simply and effectively if it satisfies Eq. (3.7) (or Eq. (3.8)).

It is obvious from Eqs. (3.14) and (3.3) that the interval-valued equal surplus division value $\overline{\rho}^{\mathrm{ESD}}(\overline{v})$ and the interval-valued equal division value $\overline{\rho}^{\mathrm{ED}}(\overline{v})$ of an interval-valued cooperative game $\overline{v} \in \overline{G}^n$ coincide if $\overline{v}(i) = 0$ for all players $i \in N$, i.e., $v_L(i) = v_R(i) = 0$ $(i = 1, 2, \ldots, n)$.

Example 3.3 Let us consider a similar cooperative production problem in which the situation is stated as in Example 1.1. We assume that the interval-valued characteristic function of the corresponding interval-valued cooperative game $\overline{v}' \in \overline{G}^3$ is changed as follows: $\overline{v}'(1) = [0, 2]$, $\overline{v}'(2) = [1/2, 3/2]$, $\overline{v}'(3) = [1, 2]$, $\overline{v}'(1, 2) = [1, 2]$, $\overline{v}'(2, 3) = [3/2, 5/2]$, $\overline{v}'(1, 3) = [1, 3]$, and $\overline{v}'(1, 2, 3) = [6, 7]$, where $N' = \{1, 2, 3\}$ and $\overline{v}'(\varnothing) = 0$. Now, we want to compute the interval-valued equal surplus division value of the interval-valued cooperative game $\overline{v}' \in \overline{G}^3$.

Using the above values (i.e., intervals) of the grand coalition N' and the individual players i $(i = 1, 2, 3)$, namely, $\overline{v}'(1) = [0, 2]$, $\overline{v}'(2) = [1/2, 3/2]$, $\overline{v}'(3) = [1, 2]$, and $\overline{v}'(1, 2, 3) = [6, 7]$, we directly have

$$v_R'\left(N'\right) - v_L'\left(N'\right) = 7 - 6 = 1,$$

$$\sum_{j=1}^{3} \left(v_R'(j) - v_L'(j)\right) = (2 - 0) + \left(\frac{3}{2} - \frac{1}{2}\right) + (2 - 1) = 4,$$

$$-3\left(v_R'(1) - v_L'(1)\right) + \sum_{j=1}^{3} \left(v_R'(j) - v_L'(j)\right) = -3 \times (2 - 0) + 4 = -2,$$

$$-3\left(v_R'(2) - v_L'(2)\right) + \sum_{j=1}^{3} \left(v_R'(j) - v_L'(j)\right) = -3 \times \left(\frac{3}{2} - \frac{1}{2}\right) + 4 = 1,$$

and

$$-3\left(v_R'(3) - v_L'(3)\right) + \sum_{j=1}^{3} \left(v_R'(j) - v_L'(j)\right) = -3 \times (2 - 1) + 4 = 1.$$

Hereby, we have

$$v'_R(N') - v'_L(N') > -3\left(v'_R(1) - v'_L(1)\right) + \sum_{j=1}^{3}\left(v'_R(j) - v'_L(j)\right),$$

$$v'_R(N') - v'_L(N') = -3\left(v'_R(2) - v'_L(2)\right) + \sum_{j=1}^{3}\left(v'_R(j) - v'_L(j)\right),$$

and

$$v'_R(N') - v'_L(N') = -3\left(v'_R(3) - v'_L(3)\right) + \sum_{j=1}^{3}\left(v'_R(j) - v'_L(j)\right),$$

i.e., the interval-valued cooperative game $\overline{v}' \in \overline{G}^3$ satisfies Eq. (3.8) (hereby Eq. (3.7)). Thus, according to Eq. (3.14), we can easily obtain the interval-valued payoffs of the players i $\left(i \in N' = \{1,2,3\}\right)$ in the interval-valued cooperative game $\overline{v}' \in \overline{G}^3$ as follows:

$$\overline{\rho}_1^{\mathrm{ESD}}(\overline{v}') = \left[v'_L(1) + \frac{v'_L(N') - \sum_{j=1}^{3} v'_L(j)}{3}, v'_R(1) + \frac{v'_R(N') - \sum_{j=1}^{3} v'_R(j)}{3}\right]$$

$$= \left[0 + \frac{6 - \left(0 + \frac{1}{2} + 1\right)}{3}, 2 + \frac{7 - \left(2 + \frac{3}{2} + 2\right)}{3}\right]$$

$$= \left[\frac{3}{2}, \frac{5}{2}\right],$$

$$\overline{\rho}_2^{\mathrm{ESD}}(\overline{v}') = \left[v'_L(2) + \frac{v'_L(N') - \sum_{j=1}^{3} v'_L(j)}{3}, v'_R(2) + \frac{v'_R(N') - \sum_{j=1}^{3} v'_R(j)}{3}\right]$$

$$= \left[\frac{1}{2} + \frac{6 - \left(0 + \frac{1}{2} + 1\right)}{3}, \frac{3}{2} + \frac{7 - \left(2 + \frac{3}{2} + 2\right)}{3}\right]$$

$$= [2, 2],$$

and

$$\overline{\rho}_3^{\mathrm{ESD}}\left(\overline{\upsilon}'\right) = \left[\upsilon'_L(3) + \frac{\upsilon'_L(N') - \sum_{j=1}^{3}\upsilon'_L(j)}{3}, \upsilon'_R(3) + \frac{\upsilon'_R(N') - \sum_{j=1}^{3}\upsilon'_R(j)}{3}\right]$$

$$= \left[1 + \frac{6 - \left(0 + \frac{1}{2} + 1\right)}{3}, 2 + \frac{7 - \left(2 + \frac{3}{2} + 2\right)}{3}\right]$$

$$= \left[\frac{5}{2}, \frac{5}{2}\right],$$

respectively. Therefore, we obtain the interval-valued equal surplus division value of the interval-valued cooperative game $\overline{\upsilon}' \in \overline{G}^3$ as follows:

$$\overline{\rho}^{\mathrm{ESD}}\left(\overline{\upsilon}'\right) = \left(\left[\frac{3}{2}, \frac{5}{2}\right], [2, 2], \left[\frac{5}{2}, \frac{5}{2}\right]\right)^{\mathrm{T}}.$$

Obviously, in Example 3.3, Eq. (3.9) is not satisfied due to

$$\upsilon'_R\left(N'\right) - \upsilon'_L\left(N'\right) = 1 < \sum_{j=1}^{3}\left(\upsilon'_R(j) - \upsilon'_L(j)\right) = 4.$$

As stated earlier, Eq. (3.7) plays an important role in the interval-valued equal surplus division value given by Eq. (3.14) (or Eqs. (3.12) and (3.13)) for any interval-valued cooperative game. In other word, if Eq. (3.7) is not satisfied, then the interval-valued equal surplus division value given by Eq. (3.14) is not always reasonable and correct.

Example 3.4 Let us consider a slightly modified version $\overline{\upsilon}'' \in \overline{G}^3$ of the interval-valued cooperative game $\overline{\upsilon}' \in \overline{G}^3$ given in Example 3.3. More specifically, the only difference between the interval-valued cooperative games $\overline{\upsilon}'' \in \overline{G}^3$ and $\overline{\upsilon}' \in \overline{G}^3$ is that $\overline{\upsilon}'\left(N'\right) = [6, 7]$ is modified as $\overline{\upsilon}''\left(N'\right) = [2, 5/2]$, where $N' = \{1, 2, 3\}$. Namely, the interval-valued characteristic function of the interval-valued cooperative game $\overline{\upsilon}'' \in \overline{G}^3$ is given as follows: $\overline{\upsilon}''\left(N'\right) = [2, 5/2]$ and $\overline{\upsilon}''(S) = \overline{\upsilon}'(S)$ for all other coalitions $S \subset N'$. We try to discuss the interval-valued equal surplus division value of the interval-valued cooperative game $\overline{\upsilon}'' \in \overline{G}^3$.

Using the above values (i.e., intervals) of the grand coalition N' and the individual players i ($i = 1, 2, 3$), namely, $\overline{\upsilon}''(1) = [0, 2]$, $\overline{\upsilon}''(2) = [1/2, 3/2]$, $\overline{\upsilon}''(3) = [1, 2]$, and $\overline{\upsilon}''\left(N'\right) = [2, 5/2]$, we directly have

$$v_R''\left(N'\right) - v_L''\left(N'\right) = \frac{5}{2} - 2 = \frac{1}{2},$$

$$\sum_{j=1}^{3}\left(v_R''(j) - v_L''(j)\right) = \sum_{j=1}^{3}\left(v_R'(j) - v_L'(j)\right) = 4,$$

$$-3\left(v_R'(1) - v_L'(1)\right) + \sum_{j=1}^{3}\left(v_R'(j) - v_L'(j)\right) = -3 \times (2 - 0) + 4 = -2,$$

$$-3\left(v_R'(2) - v_L'(2)\right) + \sum_{j=1}^{3}\left(v_R'(j) - v_L'(j)\right) = -3 \times \left(\frac{3}{2} - \frac{1}{2}\right) + 4 = 1,$$

and

$$-3\left(v_R'(3) - v_L'(3)\right) + \sum_{j=1}^{3}\left(v_R'(j) - v_L'(j)\right) = -3 \times (2 - 1) + 4 = 1.$$

Hereby, we have

$$v_R''\left(N'\right) - v_L''\left(N'\right) > -3\left(v_R'(1) - v_L'(1)\right) + \sum_{j=1}^{3}\left(v_R'(j) - v_L'(j)\right),$$

$$v_R''\left(N'\right) - v_L''\left(N'\right) < -3\left(v_R'(2) - v_L'(2)\right) + \sum_{j=1}^{3}\left(v_R'(j) - v_L'(j)\right),$$

and

$$v_R''\left(N'\right) - v_L''\left(N'\right) < -3\left(v_R'(3) - v_L'(3)\right) + \sum_{j=1}^{3}\left(v_R'(j) - v_L'(j)\right),$$

i.e., the interval-valued cooperative game $\overline{v}'' \in \overline{G}^3$ does not satisfy Eq. (3.8) (hereby Eq. (3.7)). But, if Eqs. (3.12) and (3.13) (or Eq. (3.14)) were used, then we can obtain the lower and upper bounds of the interval-valued payoffs of the players i $\left(i \in N' = \{1, 2, 3\}\right)$ in the interval-valued cooperative game $\overline{v}'' \in \overline{G}^3$ as follows:

$$\rho_{L1}^{\text{ESD}}\left(\overline{v}''\right) = v_L''(1) + \frac{v_L''(N') - \sum_{j=1}^{3} v_L''(j)}{3}$$

$$= 0 + \frac{2 - \left(0 + \frac{1}{2} + 1\right)}{3}$$

$$= \frac{1}{6},$$

$$\rho_{R1}^{\mathrm{ESD}}(\overline{v}'') = v_R''(1) + \frac{v_R''(N') - \sum_{j=1}^{3} v_R''(j)}{3}$$
$$= 2 + \frac{\frac{5}{2} - \left(2 + \frac{3}{2} + 2\right)}{3}$$
$$= 1,$$
$$\rho_{L2}^{\mathrm{ESD}}(\overline{v}'') = v_L''(2) + \frac{v_L''(N') - \sum_{j=1}^{3} v_L''(j)}{3}$$
$$= \frac{1}{2} + \frac{2 - \left(0 + \frac{1}{2} + 1\right)}{3}$$
$$= \frac{2}{3},$$
$$\rho_{R2}^{\mathrm{ESD}}(\overline{v}'') = v_R''(2) + \frac{v_R''(N') - \sum_{j=1}^{3} v_R''(j)}{3}$$
$$= \frac{3}{2} + \frac{\frac{5}{2} - \left(2 + \frac{3}{2} + 2\right)}{3}$$
$$= \frac{1}{2},$$
$$\rho_{L3}^{\mathrm{ESD}}(\overline{v}'') = v_L''(3) + \frac{v_L''(N') - \sum_{j=1}^{3} v_L''(j)}{3}$$
$$= 1 + \frac{2 - \left(0 + \frac{1}{2} + 1\right)}{3}$$
$$= \frac{7}{6},$$

and

$$\rho_{R3}^{\mathrm{ESD}}(\overline{v}'') = v_R''(3) + \frac{v_R''(N') - \sum_{j=1}^{3} v_R''(j)}{3}$$
$$= 2 + \frac{\frac{5}{2} - \left(2 + \frac{3}{2} + 2\right)}{3}$$
$$= 1,$$

respectively. Clearly, the above results are irrational due to

$$\rho_{L2}^{\text{ESD}}\left(\overline{\upsilon}''\right) = \frac{2}{3} > \rho_{R2}^{\text{ESD}}\left(\overline{\upsilon}''\right) = \frac{1}{2}$$

and

$$\rho_{L3}^{\text{ESD}}\left(\overline{\upsilon}''\right) = \frac{7}{6} > \rho_{R3}^{\text{ESD}}\left(\overline{\upsilon}''\right) = 1$$

from the notation of intervals stated as in the previous Sect. 1.3.1.

In the sequent, we study some useful and important properties of interval-valued equal surplus division values of interval-valued cooperative games.

Theorem 3.7 *(Existence and Uniqueness) For an arbitrary interval-valued cooperative game $\overline{\upsilon} \in \overline{G}^n$, if it satisfies Eq. (3.7), there always exists a unique interval-valued equal surplus division value $\overline{\boldsymbol{\rho}}^{\text{ESD}}(\overline{\upsilon})$, which is determined by Eq. (3.14).*

Proof According to Eq. (3.14), and combining with Definition 1.1, we can straightforwardly prove that Theorem 3.7 is valid.

Theorem 3.8 *(Efficiency) For any interval-valued cooperative game $\overline{\upsilon} \in \overline{G}^n$, if it satisfies Eq. (3.7), then its interval-valued equal surplus division value $\overline{\boldsymbol{\rho}}^{\text{ESD}}(\overline{\upsilon})$ satisfies the efficiency, i.e., $\sum_{i=1}^n \overline{\rho}_i^{\text{ESD}}(\overline{\upsilon}) = \overline{\upsilon}(N)$.*

Proof According to Eq. (3.14) and Definition 1.1, we have

$$\begin{aligned}
\sum_{i=1}^n \overline{\rho}_i^{\text{ESD}}(\overline{\upsilon}) &= \sum_{i=1}^n \left[\upsilon_L(i) + \frac{\upsilon_L(N) - \sum_{j=1}^n \upsilon_L(j)}{n}, \upsilon_R(i) + \frac{\upsilon_R(N) - \sum_{j=1}^n \upsilon_R(j)}{n} \right] \\
&= \left[\sum_{i=1}^n \upsilon_L(i) + \upsilon_L(N) - \sum_{j=1}^n \upsilon_L(j), \sum_{i=1}^n \upsilon_R(i) + \upsilon_R(N) - \sum_{j=1}^n \upsilon_R(j) \right] \\
&= [\upsilon_L(N), \upsilon_R(N)] \\
&= \overline{\upsilon}(N),
\end{aligned}$$

i.e.,

$$\sum_{i=1}^n \overline{\rho}_i^{\text{ESD}}(\overline{\upsilon}) = \overline{\upsilon}(N).$$

Therefore, we have completed the proof of Theorem 3.8.

Theorem 3.9 *(Additivity) For any two interval-valued cooperative games $\overline{\upsilon} \in \overline{G}^n$ and $\overline{\nu} \in \overline{G}^n$, if they satisfy Eq. (3.7), then $\overline{\rho}_i^{\text{ESD}}(\overline{\upsilon} + \overline{\nu}) = \overline{\rho}_i^{\text{ESD}}(\overline{\upsilon}) + \overline{\rho}_i^{\text{ESD}}(\overline{\nu})$ $(i = 1, 2, \ldots, n)$, i.e., $\overline{\boldsymbol{\rho}}^{\text{ESD}}(\overline{\upsilon} + \overline{\nu}) = \overline{\boldsymbol{\rho}}^{\text{ESD}}(\overline{\upsilon}) + \overline{\boldsymbol{\rho}}^{\text{ESD}}(\overline{\nu})$.*

Proof According to Eq. (3.14) and Definition 1.1, we have

$$\overline{\rho}_i^{\mathrm{ESD}}(\overline{\upsilon}+\overline{\nu}) = \left[(\upsilon_L(i)+\nu_L(i)) + \frac{(\upsilon_L(N)+\nu_L(N)) - \sum_{j=1}^{n}(\upsilon_L(j)+\nu_L(j))}{n}, (\upsilon_R(i)+\nu_R(i)) + \frac{(\upsilon_R(N)+\nu_R(N)) - \sum_{j=1}^{n}(\upsilon_R(j)+\nu_R(j))}{n}\right]$$

$$= \left[\upsilon_L(i) + \frac{\upsilon_L(N) - \sum_{j=1}^{n}\upsilon_L(j)}{n}, \upsilon_R(i) + \frac{\upsilon_R(N) - \sum_{j=1}^{n}\upsilon_R(j)}{n}\right] + \left[\nu_L(i) + \frac{\nu_L(N) - \sum_{j=1}^{n}\nu_L(j)}{n}, \nu_R(i) + \frac{\nu_R(N) - \sum_{j=1}^{n}\nu_R(j)}{n}\right]$$

$$= \overline{\rho}_i^{\mathrm{ESD}}(\overline{\upsilon}) + \overline{\rho}_i^{\mathrm{ESD}}(\overline{\nu}),$$

i.e.,

$$\overline{\rho}_i^{\mathrm{ESD}}(\overline{\upsilon}+\overline{\nu}) = \overline{\rho}_i^{\mathrm{ESD}}(\overline{\upsilon}) + \overline{\rho}_i^{\mathrm{ESD}}(\overline{\nu}) \quad (i = 1, 2, \ldots, n).$$

Thus, we obtain

$$\overline{\boldsymbol{\rho}}^{\mathrm{ESD}}(\overline{\upsilon}+\overline{\nu}) = \overline{\boldsymbol{\rho}}^{\mathrm{ESD}}(\overline{\upsilon}) + \overline{\boldsymbol{\rho}}^{\mathrm{ESD}}(\overline{\nu}).$$

Therefore, we have completed the proof of Theorem 3.9.

Theorem 3.10 *(Symmetry) For any interval-valued cooperative game* $\overline{\upsilon} \in \overline{G}^n$, *if it satisfies Eq. (3.7), and players* $i \in N$ *and* $k \in N$ $(i \neq k)$ *are symmetric in the interval-valued cooperative game* $\overline{\upsilon}$, *then* $\overline{\rho}_i^{\mathrm{ESD}}(\overline{\upsilon}) = \overline{\rho}_k^{\mathrm{ESD}}(\overline{\upsilon})$.

Proof Due to the assumption that the players $i \in N$ and $k \in N$ $(i \neq k)$ are symmetric in the interval-valued cooperative game $\overline{\upsilon} \in \overline{G}^n$, then according to Definition 1.3, we have

$$\overline{\upsilon}(S \cup i) = \overline{\upsilon}(S \cup k)$$

for any coalition $S \subseteq N \backslash \{i, k\}$. Particularly, we have $\overline{\upsilon}(i) = \overline{\upsilon}(k)$, i.e., $\upsilon_L(i) = \upsilon_L(k)$ and $\upsilon_R(i) = \upsilon_R(k)$. According to Eq. (3.14), we can easily obtain that

$$\overline{\rho}_i^{\mathrm{ESD}}(\overline{v}) = \left[v_L(i) + \frac{v_L(N) - \sum_{j=1}^{n} v_L(j)}{n}, v_R(i) + \frac{v_R(N) - \sum_{j=1}^{n} v_R(j)}{n} \right]$$

$$= \left[v_L(k) + \frac{v_L(N) - \sum_{j=1}^{n} v_L(j)}{n}, v_R(k) + \frac{v_R(N) - \sum_{j=1}^{n} v_R(j)}{n} \right]$$

$$= \overline{\rho}_k^{\mathrm{ESD}}(\overline{v}),$$

i.e.,

$$\overline{\rho}_i^{\mathrm{ESD}}(\overline{v}) = \overline{\rho}_k^{\mathrm{ESD}}(\overline{v}).$$

Thus, we have completed the proof of Theorem 3.10.

Theorem 3.11 *(Anonymity) For an arbitrary interval-valued cooperative game* $\overline{v} \in \overline{G}^n$ *and any permutation* σ *on the set* N*, if* $\overline{v}$ *satisfies Eq. (3.7), then* $\overline{\rho}_{\sigma(i)}^{\mathrm{ESD}}(\overline{v}^\sigma) = \overline{\rho}_i^{\mathrm{ESD}}(\overline{v})$ $(i = 1, 2, \ldots, n)$*. Namely,* $\overline{\boldsymbol{\rho}}^{\mathrm{ESD}}(\overline{v}^\sigma) = \boldsymbol{\sigma}^{\#}\left(\overline{\boldsymbol{\rho}}^{\mathrm{ESD}}(\overline{v})\right)$*.*

Proof According to Eq. (3.14), we can easily prove Theorem 3.11 in a similar way to that of Theorem 3.10 (omitted).

For any real number $a \in R$, we define a new interval-valued cooperative game $\overline{\nu} \in \overline{G}^n$ associated with the interval-valued cooperative game $\overline{v} \in \overline{G}^n$ as follows:

$$\overline{\nu}(S) = a\overline{v}(S) + \sum_{i \in S} \overline{d}_i \quad (S \subseteq N), \tag{3.15}$$

where $\overline{d}_i \in \overline{R}$ $(i \in N)$ is an interval. Denote $\overline{\boldsymbol{d}} = \left(\overline{d}_1, \overline{d}_2, \ldots, \overline{d}_n\right)^{\mathrm{T}} \in \overline{R}^n$.

Theorem 3.12 *(Invariance) For any interval-valued cooperative game* $\overline{v} \in \overline{G}^n$ *and its associated interval-valued cooperative game* $\overline{\nu} \in \overline{G}^n$ *given by Eq. (3.15), if they satisfy Eq. (3.7), then* $\overline{\rho}_i^{\mathrm{ESD}}(\overline{\nu}) = a\overline{\rho}_i^{\mathrm{ESD}}(\overline{v}) + \overline{d}_i$ $(i = 1, 2, \ldots, n)$*, i.e.,* $\overline{\boldsymbol{\rho}}^{\mathrm{ESD}}(\overline{\nu}) = a\overline{\boldsymbol{\rho}}^{\mathrm{ESD}}(\overline{v}) + \overline{\boldsymbol{d}}$*.*

Proof According to Eq. (3.14) and Definition 1.1, we have

$$\overline{\rho}_i^{\mathrm{ESD}}(\overline{v}) = \left[(av_L(i)+d_{Li}) + \frac{\left(av_L(N)+\sum_{i=1}^{n} d_{Li}\right) - \sum_{j=1}^{n}(av_L(j)+d_{Lj})}{n}, (av_R(i)+d_{Ri}) + \frac{\left(av_R(N)+\sum_{i=1}^{n} d_{Ri}\right) - \sum_{j=1}^{n}(av_R(j)+d_{Rj})}{n}\right]$$

$$= \left[av_L(i) + \frac{av_L(N) - \sum_{j=1}^{n} av_L(j)}{n}, av_R(i) + \frac{av_R(N) - \sum_{j=1}^{n} av_R(j)}{n}\right] + [d_{Li}, d_{Ri}]$$

$$= a\left[v_L(i) + \frac{v_L(N) - \sum_{j=1}^{n} v_L(j)}{n}, v_R(i) + \frac{v_R(N) - \sum_{j=1}^{n} v_R(j)}{n}\right] + [d_{Li}, d_{Ri}]$$

$$= a\overline{\rho}_i^{\mathrm{ESD}}(\overline{v}) + \overline{d}_i,$$

i.e.,

$$\overline{\rho}_i^{\mathrm{ESD}}(\overline{v}) = a\overline{\rho}_i^{\mathrm{ESD}}(\overline{v}) + \overline{d}_i \quad (i = 1, 2, \ldots, n).$$

Hereby, we obtain

$$\overline{\boldsymbol{\rho}}^{\mathrm{ESD}}(\overline{v}) = a\overline{\boldsymbol{\rho}}^{\mathrm{ESD}}(\overline{v}) + \overline{\boldsymbol{d}}.$$

Thus, we have completed the proof of Theorem 3.12.

Note that interval-valued equal surplus division values of interval-valued cooperative games always satisfy the invariance whereas interval-valued equal division values do not satisfy the invariance.

From Eq. (3.14), obviously, interval-valued equal surplus division values of interval-valued cooperative games do not satisfy the dummy player property and the null player property.

Furthermore, interval-valued equal surplus division values of interval-valued cooperative games do not always satisfy the individual rationality.

Example 3.5 Let us consider other modified version $\overline{v}''' \in \overline{G}^3$ of the interval-valued cooperative game $\overline{v}' \in \overline{G}^3$ given in Example 3.3. To be more specific, the interval-valued characteristic function of the interval-valued cooperative game $\overline{v}''' \in \overline{G}^3$ is given as follows: $\overline{v}'''(N') = [2, 3]$ and $\overline{v}'''(S) = \overline{v}'(S)$ for all other coalitions $S \subset N'$, where $N' = \{1, 2, 3\}$. In other word, the only difference between the interval-valued cooperative games $\bar{v}''' \in \bar{G}^3$ and $\bar{v}' \in \bar{G}^3$ is that $\bar{v}'(N') = [6, 7]$ is modified as $\bar{v}'''(N') = [2, 3]$. We discuss the individual rationality of the interval-valued equal surplus division value of the interval-valued cooperative game $\bar{v}''' \in \bar{G}^3$.

In a similar way to that of Example 3.3, we directly have

$$v_R'''\left(N'\right) - v_L'''\left(N'\right) = 3 - 2 = 1,$$

$$\sum_{j=1}^{3}\left(v_R'''(j) - v_L'''(j)\right) = \sum_{j=1}^{3}\left(v_R'(j) - v_L'(j)\right) = 4,$$

$$-3\left(v_R'(1) - v_L'(1)\right) + \sum_{j=1}^{3}\left(v_R'(j) - v_L'(j)\right) = -3 \times (2 - 0) + 4 = -2,$$

$$-3\left(v_R'(2) - v_L'(2)\right) + \sum_{j=1}^{3}\left(v_R'(j) - v_L'(j)\right) = -3 \times \left(\frac{3}{2} - \frac{1}{2}\right) + 4 = 1,$$

and

$$-3\left(v_R'(3) - v_L'(3)\right) + \sum_{j=1}^{3}\left(v_R'(j) - v_L'(j)\right) = -3 \times (2 - 1) + 4 = 1.$$

Hereby, we have

$$v_R'''\left(N'\right) - v_L'''\left(N'\right) > -3\left(v_R'(1) - v_L'(1)\right) + \sum_{j=1}^{3}\left(v_R'(j) - v_L'(j)\right),$$

$$v_R'''\left(N'\right) - v_L'''\left(N'\right) = -3\left(v_R'(2) - v_L'(2)\right) + \sum_{j=1}^{3}\left(v_R'(j) - v_L'(j)\right),$$

and

$$v_R'''\left(N'\right) - v_L'''\left(N'\right) = -3\left(v_R'(3) - v_L'(3)\right) + \sum_{j=1}^{3}\left(v_R'(j) - v_L'(j)\right),$$

i.e., the interval-valued cooperative game $\bar{v}''' \in \bar{G}^3$ satisfies Eq. (3.8) (hereby Eq. (3.7)). Thus, using Eqs. (3.12) and (3.13), we can obtain the lower and upper bounds of the components of the interval-valued equal surplus division value of the interval-valued cooperative game $\bar{v}''' \in \bar{G}^3$ as follows:

$$\begin{aligned}\rho_{L1}^{\mathrm{ESD}}\left(\bar{v}'''\right) &= \bar{v}_L'''(1) + \frac{\bar{v}_L'''\left(N'\right) - \sum_{j=1}^{3}\bar{v}_L'''(j)}{3}\\ &= 0 + \frac{2 - \left(0 + \frac{1}{2} + 1\right)}{3}\\ &= \frac{1}{6},\end{aligned}$$

$$\rho_{R1}^{\mathrm{ESD}}(\overline{v}''') = \overline{v}_R'''(1) + \frac{\overline{v}_R'''(N') - \sum_{j=1}^{3}\overline{v}_R'''(j)}{3}$$

$$= 2 + \frac{3 - \left(2 + \frac{3}{2} + 2\right)}{3}$$

$$= \frac{7}{6},$$

$$\rho_{L2}^{\mathrm{ESD}}(\overline{v}''') = \overline{v}_L'''(2) + \frac{\overline{v}_L'''(N') - \sum_{j=1}^{3}\overline{v}_L'''(j)}{3}$$

$$= \frac{1}{2} + \frac{2 - \left(0 + \frac{1}{2} + 1\right)}{3}$$

$$= \frac{2}{3},$$

$$\rho_{R2}^{\mathrm{ESD}}(\overline{v}''') = \overline{v}_R'''(2) + \frac{\overline{v}_R'''(N') - \sum_{j=1}^{3}\overline{v}_R'''(j)}{3}$$

$$= \frac{3}{2} + \frac{3 - \left(2 + \frac{3}{2} + 2\right)}{3}$$

$$= \frac{2}{3},$$

$$\rho_{L3}^{\mathrm{ESD}}(\overline{v}''') = \overline{v}_L'''(3) + \frac{\overline{v}_L'''(N') - \sum_{j=1}^{3}\overline{v}_L'''(j)}{3}$$

$$= 1 + \frac{2 - \left(0 + \frac{1}{2} + 1\right)}{3}$$

$$= \frac{7}{6},$$

and

$$\rho_{R3}^{\mathrm{ESD}}(\overline{v}''') = \overline{v}_R'''(3) + \frac{\overline{v}_R'''(N') - \sum_{j=1}^{3}\overline{v}_R'''(j)}{3}$$

$$= 2 + \frac{3 - \left(2 + \frac{3}{2} + 2\right)}{3}$$

$$= \frac{7}{6}.$$

Namely, $\overline{\rho}_1^{\text{ESD}}(\overline{\upsilon}''') = [1/6, 7/6]$, $\overline{\rho}_2^{\text{ESD}}(\overline{\upsilon}''') = [2/3, 2/3]$, and $\overline{\rho}_3^{\text{ESD}}(\overline{\upsilon}''') = [7/6, 7/6]$. Obviously, in the Moore's order relation or ranking for intervals [5], i.e., Eq. (1.4), we have

$$\rho_{R1}^{\text{ESD}}\left(\overline{\upsilon}'''\right) = \frac{7}{6} < 2 = \upsilon_R'''(1),$$

$$\rho_{R2}^{\text{ESD}}\left(\overline{\upsilon}'''\right) = \frac{2}{3} < \frac{3}{2} = \upsilon_R'''(2),$$

and

$$\rho_{R3}^{\text{ESD}}\left(\overline{\upsilon}'''\right) = \frac{7}{6} < 2 = \upsilon_R'''(3).$$

That is to say, the interval-valued equal surplus division value $\overline{\boldsymbol{\rho}}^{\text{ESD}}\left(\overline{\upsilon}'''\right)$ of the interval-valued cooperative game $\overline{\upsilon}''' \in \overline{G}^3$ does not satisfy the individual rationality. The reason is that

$$\upsilon_R'''\left(N'\right) = 3 < \sum_{j=1}^{3} \upsilon_R'''(j) = 2 + \frac{3}{2} + 2 = \frac{11}{2},$$

i.e., the interval-valued cooperative game $\overline{\upsilon}''' \in \overline{G}^3$ does not satisfy Eq. (3.10) although Eq. (3.11) is satisfied.

However, for an interval-valued cooperative game $\overline{\upsilon} \in \overline{G}^n$ which satisfies Eq. (3.7) (or Eq. (3.9)), if it also satisfies Eqs. (3.10) and (3.11) at the same time, then its interval-valued equal surplus division value $\overline{\boldsymbol{\rho}}^{\text{ESD}}(\overline{\upsilon})$ satisfies the individual rationality.

3.3 Interval-Valued Egalitarian Shapley Values and Interval-Valued Discounted Shapley Values of Interval-Valued Cooperative Games

Stated as in the previous Sect. 1.2, the Shapley value is an important and a commonly used single-valued solution concept in cooperative game theory [16, 17]. In cooperative games under interval uncertainty, some researchers pay attention to the Shapley value. Particularly, for some special interval-valued cooperative games, extended Shapley values such as the interval-valued Shapley-like value were developed on the basis of the special interval subtraction (e.g., Hukuhara difference [14]) or through imposing some special constraint conditions [18–20]. In this section, inspired by Li [21, 22], and in the same way to the above

Sect. 3.2, we will develop effective and simplified methods for computing the interval-valued Shapley value and its extensions and/or variants of interval-valued cooperative games through using the monotonicity rather than the special interval subtraction operator or ranking method of intervals.

3.3.1 Interval-Valued Shapley Values of Interval-Valued Cooperative Games and Simplified Methods

Let us continue to consider the interval-valued cooperative game $\overline{v} \in \overline{G}^n$ given in Sect. 1.3.2.

In the same way to the above Sect. 3.2, we can construct an associated cooperative game $v(\alpha) \in G^n$, where the set of players is $N = \{1, 2, \ldots, n\}$ and the characteristic function $v(\alpha)$ of coalitions of players is given by Eq. (3.4).

According to Eq. (1.3), we can easily obtain the Shapley value $\boldsymbol{\Phi}^{\mathrm{SH}}(v(\alpha)) = (\phi_1^{\mathrm{SH}}(v(\alpha)), \phi_2^{\mathrm{SH}}(v(\alpha)), \ldots, \phi_n^{\mathrm{SH}}(v(\alpha)))^{\mathrm{T}}$ of the cooperative game $v(\alpha) \in G^n$, where

$$\phi_i^{\mathrm{SH}}(v(\alpha)) = \sum_{S \subseteq N \backslash i} \frac{s!(n-s-1)!}{n!}(v(\alpha)(S \cup i) - v(\alpha)(S)) \quad (i = 1, 2, \ldots, n),$$

which can further be rewritten as follows:

$$\begin{aligned}\phi_i^{\mathrm{SH}}(v(\alpha)) = \sum_{S \subseteq N \backslash i} \Bigg\{ & \frac{s!(n-s-1)!}{n!} \\ & \times \{[(1-\alpha)v_L(S \cup i) + \alpha v_R(S \cup i)] - [(1-\alpha)v_L(S) + \alpha v_R(S)]\} \Bigg\} \\ & (i = 1, 2, \ldots, n),\end{aligned} \tag{3.16}$$

where $\alpha \in [0, 1]$.

Obviously, the Shapley value $\boldsymbol{\Phi}^{\mathrm{SH}}(v(\alpha))$ is a continuous function of the characteristic function value $v(\alpha)$ of coalitions in N. Note that $v(\alpha)$ is also a continuous function of $\alpha \in [0, 1]$ due to Eq. (3.4). Accordingly, the Shapley value $\boldsymbol{\Phi}^{\mathrm{SH}}(v(\alpha))$ is a continuous function of the parameter $\alpha \in [0, 1]$.

Theorem 3.13 *For any interval-valued cooperative game $\overline{v} \in \overline{G}^n$, if the following system of inequalities*

$$v_R(S \cup i) - v_L(S \cup i) \geq v_R(S) - v_L(S) \quad (i = 1, 2, \ldots, n; S \subseteq N \backslash i) \tag{3.17}$$

is satisfied, then the Shapley value $\boldsymbol{\Phi}^{\mathrm{SH}}(v(\alpha))$ of the cooperative game $v(\alpha) \in G^n$ is a monotonic and non-decreasing function of the parameter $\alpha \in [0, 1]$.

Proof For any $\alpha \in [0,1]$ and $\alpha' \in [0,1]$, according to Eq. (3.16), and using Eq. (3.4), we have

$$\begin{aligned}\phi_i^{\mathrm{SH}}(v(\alpha)) - \phi_i^{\mathrm{SH}}(v(\alpha')) &= \sum_{S \subseteq N \setminus i} \left\{ \frac{s!(n-s-1)!}{n!} \left[\left(v(\alpha)(S \cup i) - v\left(\alpha'\right)(S \cup i) \right) \right.\right. \\ &\quad \left.\left. - \left(v(\alpha)(S) - v(\alpha')(S) \right) \right] \right\} \\ &= \left(\alpha - \alpha'\right) \sum_{S \subseteq N \setminus i} \left\{ \frac{s!(n-s-1)!}{n!} \right. \\ &\quad \left. \times \left[\left(v_R(S \cup i) - v_L(S \cup i) \right) - \left(v_R(S) - v_L(S) \right) \right] \right\},\end{aligned}$$

where $i = 1, 2, \ldots, n$.

If $\alpha \geq \alpha'$, then combining with the assumption, i.e., Eq. (3.17), we have

$$\phi_i^{\mathrm{SH}}(v(\alpha)) - \phi_i^{\mathrm{SH}}\left(v\left(\alpha'\right)\right) \geq 0 \quad (i = 1, 2, \ldots, n),$$

i.e.,

$$\phi_i^{\mathrm{SH}}(v(\alpha)) \geq \phi_i^{\mathrm{SH}}\left(v\left(\alpha'\right)\right) \quad (i = 1, 2, \ldots, n),$$

which mean that the Shapley value $\boldsymbol{\Phi}^{\mathrm{SH}}(v(\alpha))$ is a monotonic and non-decreasing function of the parameter $\alpha \in [0,1]$. Thus, we have completed the proof of Theorem 3.13.

Therefore, for any interval-valued cooperative game $\overline{v} \in \overline{G}^n$, if it satisfies Eq. (3.17), then it is directly derived from Theorem 3.13 and Eq. (3.16) that the lower and upper bounds of the components (intervals) $\overline{\phi}_i^{\mathrm{SH}}(\overline{v})$ $(i = 1, 2, \ldots, n)$ of the interval-valued Shapley value $\overline{\boldsymbol{\Phi}}^{\mathrm{SH}}(\overline{v}) = \left(\overline{\phi}_1^{\mathrm{SH}}(\overline{v}), \overline{\phi}_2^{\mathrm{SH}}(\overline{v}), \ldots, \overline{\phi}_n^{\mathrm{SH}}(\overline{v}) \right)^{\mathrm{T}}$ are given as follows:

$$\phi_{Li}^{\mathrm{SH}}(\overline{v}) = \phi_i^{\mathrm{SH}}(v(0)) = \sum_{S \subseteq N \setminus i} \frac{s!(n-s-1)!}{n!} \left(v_L(S \cup i) - v_L(S) \right) \quad (i = 1, 2, \ldots, n)$$

and

$$\phi_{Ri}^{\mathrm{SH}}(\overline{v}) = \phi_i^{\mathrm{SH}}(v(1)) = \sum_{S \subseteq N \setminus i} \frac{s!(n-s-1)!}{n!} \left(v_R(S \cup i) - v_R(S) \right) \quad (i = 1, 2, \ldots, n),$$

respectively, i.e.,

$$\phi_{Li}^{\mathrm{SH}}(\bar{v}) = \sum_{S \subseteq N \backslash i} \frac{s!(n-s-1)!}{n!}(v_L(S \cup i) - v_L(S)) \quad (i = 1, 2, \ldots, n) \tag{3.18}$$

and

$$\phi_{Ri}^{\mathrm{SH}}(\bar{v}) = \sum_{S \subseteq N \backslash i} \frac{s!(n-s-1)!}{n!}(v_R(S \cup i) - v_R(S)) \quad (i = 1, 2, \ldots, n). \tag{3.19}$$

Thus, the interval-valued Shapley values $\bar{\phi}_i^{\mathrm{SH}}(\bar{v}) = \left[\phi_{Li}^{\mathrm{SH}}(\bar{v}), \phi_{Ri}^{\mathrm{SH}}(\bar{v})\right]$ of the players i $(i = 1, 2, \ldots, n)$ in the interval-valued cooperative game $\bar{v} \in \bar{G}^n$ are directly and explicitly expressed as follows:

$$\bar{\phi}_i^{\mathrm{SH}}(\bar{v}) = \left[\sum_{S \subseteq N \backslash i} \frac{s!(n-s-1)!}{n!}(v_L(S \cup i) - v_L(S)), \sum_{S \subseteq N \backslash i} \frac{s!(n-s-1)!}{n!}(v_R(S \cup i) - v_R(S))\right]$$
$$(i = 1, 2, \ldots, n). \tag{3.20}$$

Equation (3.17) is an important condition which ensures that the Shapley value $\boldsymbol{\Phi}^{\mathrm{SH}}(v(\alpha))$ possesses the monotonicity. Equation (3.17) requires that the length cooperative game $l \in G^n$ associated with the interval-valued cooperative game $\bar{v} \in \bar{G}^n$ is monotonic, i.e.,

$$l(S \cup i) \geq l(S) \quad (i = 1, 2, \ldots, n; S \subseteq N \backslash i),$$

where the characteristic function of the length cooperative game $l \in G^n$ is given as follows:

$$l(S) = v_R(S) - v_L(S) \quad (S \subseteq N \backslash i).$$

Thus, if an interval-valued cooperative game $\bar{v} \in \bar{G}^n$ satisfies Eq. (3.17), then it is called size monotonic. In a size monotonic interval-valued cooperative game, the uncertainty on coalition values does not decrease when the coalitions grow larger.

Example 3.6 The economic situation is stated as in Example 1.1. We change the interval-valued characteristic function of the interval-valued cooperative game $\bar{v}' \in \bar{G}^3$ given in Example 1.1 and hereby construct a new interval-valued cooperative game $\bar{v}^0 \in \bar{G}^3$, where $N' = \{1, 2, 3\}$ and the interval-valued characteristic function of the interval-valued cooperative game $\bar{v}^0 \in \bar{G}^3$ is given as follows: $\bar{v}^0(1) = [0, 1]$, $\bar{v}^0(2) = [1, 3]$, $\bar{v}^0(3) = [2, 4]$, $\bar{v}^0(1, 2) = [2, 4]$, $\bar{v}^0(1, 3) = [1, 4]$, $\bar{v}^0(2, 3) = [2, 5]$, $\bar{v}^0(N') = [2, 7]$, and $\bar{v}^0(\varnothing) = 0$. Let us compute the interval-valued Shapley value of the interval-valued cooperative game $\bar{v}^0 \in \bar{G}^3$.

Obviously, the interval-valued cooperative game $\overline{v}^0 \in \overline{G}^3$ satisfies Eq. (3.17), i.e., its associated length cooperative game $l^0 \in G^3$ is monotonic. Thus, according to Eqs. (3.18) and (3.19) (or Eq. (3.20)), we can obtain

$$\begin{aligned}
\phi_{L1}^{SH}(\overline{v}^0) &= \sum_{S\subseteq\{2,3\}} \frac{s!(2-s)!}{3!}\left(\overline{v}_L^0(S\cup 1) - \overline{v}_L^0(S)\right) \\
&= \frac{0!2!}{3!}\left(\overline{v}_L^0(1) - \overline{v}_L^0(\varnothing)\right) + \frac{1!1!}{3!}\left(\overline{v}_L^0(1,2) - \overline{v}_L^0(2)\right) + \frac{1!1!}{3!}\left(\overline{v}_L^0(1,3) - \overline{v}_L^0(3)\right) \\
&\quad + \frac{2!0!}{3!}\left(\overline{v}_L^0\left(N'\right) - \overline{v}_L^0(2,3)\right) \\
&= \frac{1}{3}(0-0) + \frac{1}{6}(2-1) + \frac{1}{6}(1-2) + \frac{1}{3}(2-2) \\
&= 0,
\end{aligned}$$

$$\begin{aligned}
\phi_{R1}^{SH}(\overline{v}^0) &= \sum_{S\subseteq\{2,3\}} \frac{s!(2-s)!}{3!}\left(\overline{v}_R^0(S\cup 1) - \overline{v}_R^0(S)\right) \\
&= \frac{0!2!}{3!}\left(\overline{v}_R^0(1) - \overline{v}_R^0(\varnothing)\right) + \frac{1!1!}{3!}\left(\overline{v}_R^0(1,2) - \overline{v}_R^0(2)\right) + \frac{1!1!}{3!}\left(\overline{v}_R^0(1,3) - \overline{v}_R^0(3)\right) \\
&\quad + \frac{2!0!}{3!}\left(\overline{v}_R^0\left(N'\right) - \overline{v}_R^0(2,3)\right) \\
&= \frac{1}{3}(1-0) + \frac{1}{6}(4-3) + \frac{1}{6}(4-4) + \frac{1}{3}(7-5) \\
&= \frac{7}{6},
\end{aligned}$$

$$\begin{aligned}
\phi_{L2}^{SH}(\overline{v}^0) &= \sum_{S\subseteq\{1,3\}} \frac{s!(2-s)!}{3!}\left(\overline{v}_L^0(S\cup 2) - \overline{v}_L^0(S)\right) \\
&= \frac{0!2!}{3!}\left(\overline{v}_L^0(2) - \overline{v}_L^0(\varnothing)\right) + \frac{1!1!}{3!}\left(\overline{v}_L^0(1,2) - \overline{v}_L^0(1)\right) + \frac{1!1!}{3!}\left(\overline{v}_L^0(2,3) - \overline{v}_L^0(3)\right) \\
&\quad + \frac{2!0!}{3!}\left(\overline{v}_L^0\left(N'\right) - \overline{v}_L^0(1,3)\right) \\
&= \frac{1}{3}(1-0) + \frac{1}{6}(2-0) + \frac{1}{6}(2-2) + \frac{1}{3}(2-1) \\
&= 1,
\end{aligned}$$

$$\begin{aligned}
\phi_{R2}^{\text{SH}}(\overline{v}^0) &= \sum_{S \subseteq \{1,3\}} \frac{s!(2-s)!}{3!}\left(\overline{v}_R^0(S \cup 2) - \overline{v}_R^0(S)\right) \\
&= \frac{0!2!}{3!}\left(\overline{v}_R^0(2) - \overline{v}_R^0(\varnothing)\right) + \frac{1!1!}{3!}\left(\overline{v}_R^0(1,2) - \overline{v}_R^0(1)\right) + \frac{1!1!}{3!}\left(\overline{v}_R^0(2,3) - \overline{v}_R^0(3)\right) \\
&\quad + \frac{2!0!}{3!}\left(\overline{v}_R^0\left(N'\right) - \overline{v}_R^0(1,3)\right) \\
&= \frac{1}{3}(3-0) + \frac{1}{6}(4-1) + \frac{1}{6}(5-4) + \frac{1}{3}(7-4) \\
&= \frac{8}{3},
\end{aligned}$$

$$\begin{aligned}
\phi_{L3}^{\text{SH}}(\overline{v}^0) &= \sum_{S \subseteq \{1,2\}} \frac{s!(2-s)!}{3!}\left(\overline{v}_L^0(S \cup 3) - \overline{v}_L^0(S)\right) \\
&= \frac{0!2!}{3!}\left(\overline{v}_L^0(3) - \overline{v}_L^0(\varnothing)\right) + \frac{1!1!}{3!}\left(\overline{v}_L^0(1,3) - \overline{v}_L^0(1)\right) + \frac{1!1!}{3!}\left(\overline{v}_L^0(2,3) - \overline{v}_L^0(2)\right) \\
&\quad + \frac{2!0!}{3!}\left(\overline{v}_L^0\left(N'\right) - \overline{v}_L^0(1,2)\right) \\
&= \frac{1}{3}(2-0) + \frac{1}{6}(1-0) + \frac{1}{6}(2-1) + \frac{1}{3}(2-2) \\
&= 1,
\end{aligned}$$

and

$$\begin{aligned}
\phi_{R3}^{\text{SH}}(\overline{v}^0) &= \sum_{S \subseteq \{1,2\}} \frac{s!(2-s)!}{3!}\left(\overline{v}_R^0(S \cup 3) - \overline{v}_R^0(S)\right) \\
&= \frac{0!2!}{3!}\left(\overline{v}_R^0(3) - \overline{v}_R^0(\varnothing)\right) + \frac{1!1!}{3!}\left(\overline{v}_R^0(1,3) - \overline{v}_R^0(1)\right) + \frac{1!1!}{3!}\left(\overline{v}_R^0(2,3) - \overline{v}_R^0(2)\right) \\
&\quad + \frac{2!0!}{3!}\left(\overline{v}_R^0\left(N'\right) - \overline{v}_R^0(1,2)\right) \\
&= \frac{1}{3}(4-0) + \frac{1}{6}(4-1) + \frac{1}{6}(5-3) + \frac{1}{3}(7-4) \\
&= \frac{19}{6},
\end{aligned}$$

respectively. Hence, we can obtain the interval-valued Shapley value of the interval-valued cooperative game $\overline{v}^0 \in \overline{G}^3$ as follows:

$$\overline{\boldsymbol{\Phi}}^{\text{SH}}(\overline{v}^0) = \left(\left[0, \frac{7}{6}\right], \left[1, \frac{8}{3}\right], \left[1, \frac{19}{6}\right]\right)^{\text{T}}.$$

Particularly, for any two-person interval-valued cooperative game $\overline{v} \in \overline{G}^2$, if it satisfies Eq. (3.17), then according to Eqs. (3.18) and (3.19), we can explicitly rewrite the lower and upper bounds of the interval-valued Shapley values $\overline{\phi}_i^{\text{SH}}(\overline{v})$ of the players i $(i = 1, 2)$ in the interval-valued cooperative game $\overline{v} \in \overline{G}^2$ as follows:

$$\phi_{L1}^{\text{SH}}(\overline{v}) = \frac{v_L(1) + v_L(1,2) - v_L(2)}{2}, \tag{3.21}$$

$$\phi_{R1}^{\text{SH}}(\overline{v}) = \frac{v_R(1) + v_R(1,2) - v_R(2)}{2}, \tag{3.22}$$

$$\phi_{L2}^{\text{SH}}(\overline{v}) = \frac{v_L(2) + v_L(1,2) - v_L(1)}{2}, \tag{3.23}$$

and

$$\phi_{R2}^{\text{SH}}(\overline{v}) = \frac{v_R(2) + v_R(1,2) - v_R(1)}{2}. \tag{3.24}$$

Therefore, we can simply rewrite the interval-valued Shapley values $\overline{\phi}_i^{\text{SH}}(\overline{v})$ of the players i $(i = 1, 2)$ in the interval-valued cooperative game $\overline{v} \in \overline{G}^2$ as follows:

$$\overline{\phi}_1^{\text{SH}}(\overline{v}) = \left[\frac{v_L(1) + v_L(1,2) - v_L(2)}{2}, \frac{v_R(1) + v_R(1,2) - v_R(2)}{2}\right] \tag{3.25}$$

and

$$\overline{\phi}_2^{\text{SH}}(\overline{v}) = \left[\frac{v_L(2) + v_L(1,2) - v_L(1)}{2}, \frac{v_R(2) + v_R(1,2) - v_R(1)}{2}\right]. \tag{3.26}$$

Alternatively, Eqs. (3.25) and (3.26) can be rewritten as follows:

$$\overline{\phi}_1^{\text{SH}}(\overline{v}) = \left[v_L(1) + \frac{v_L(1,2) - (v_L(1) + v_L(2))}{2}, v_R(1) + \frac{v_R(1,2) - (v_R(1) + v_R(2))}{2}\right]$$

and

$$\overline{\phi}_2^{\text{SH}}(\overline{v}) = \left[v_L(2) + \frac{v_L(1,2) - (v_L(1) + v_L(2))}{2}, v_R(2) + \frac{v_R(1,2) - (v_R(1) + v_R(2))}{2}\right],$$

which is just about the interval-valued equal surplus division value given by Eq. (3.14) when $n = 2$.

Example 3.7 Let us discuss the interval-valued Shapley value of the interval-valued cooperative game $\overline{v}'' \in \overline{G}^2$ given in Example 1.2.

From Example 1.2, we have $\bar{v}''(2) = [2,5]$ and $\bar{v}''(1,2) = [4,6]$. Therefore, we obtain

$$\bar{v}''_R(1,2) - \bar{v}''_L(1,2) = 6 - 4 = 2 < \bar{v}''_R(2) - \bar{v}''_L(2) = 5 - 2 = 3.$$

Thus, the interval-valued cooperative game $\bar{v}'' \in \overline{G}^2$ does not satisfy Eq. (3.17). In this case, if Eqs. (3.21) and (3.22) were used, then we have

$$\phi_{L1}^{SH}(\bar{v}) = \frac{4 + 0.3 - 2}{2} = 1.15$$

and

$$\phi_{R1}^{SH}(\bar{v}) = \frac{6 + 1 - 5}{2} = 1.$$

Clearly,

$$\phi_{L1}^{SH}(\bar{v}) = 1.15 > \phi_{R1}^{SH}(\bar{v}) = 1,$$

which conflicts with the notation of intervals given in the previous Sect. 1.3.1. Therefore, the interval-valued cooperative game $\bar{v}'' \in \overline{G}^2$ has not the interval-valued Shapley value defined by Eq. (3.20) (or Eqs. (3.18) and (3.19)). The reason is that the interval-valued cooperative game $\bar{v}'' \in \overline{G}^2$ is not size monotonic or the length cooperative game associated with the interval-valued cooperative game $\bar{v}'' \in \overline{G}^2$ is not monotonic.

In the sequent, we discuss some useful and important properties of interval-valued Shapley values of interval-valued cooperative games.

Theorem 3.14 *(Existence and Uniqueness) For an arbitrary interval-valued cooperative game $\bar{v} \in \overline{G}^n$, if it satisfies Eq. (3.17), there always exists a unique interval-valued Shapley value $\overline{\boldsymbol{\Phi}}^{SH}(\bar{v})$, which is determined by Eq. (3.20).*

Proof According to Eq. (3.20), and combining with Definition 1.1, we can easily prove Theorem 3.14.

Theorem 3.15 *(Efficiency) For any interval-valued cooperative game $\bar{v} \in \overline{G}^n$, if it satisfies Eq. (3.17), then its interval-valued Shapley value $\overline{\boldsymbol{\Phi}}^{SH}(\bar{v})$ satisfies the efficiency, i.e., $\sum_{i=1}^n \bar{\phi}_i^{SH}(\bar{v}) = \bar{v}(N)$.*

Proof According to Eqs. (3.18) and (3.19), we have

$$\sum_{i=1}^{n} \phi_{Li}^{SH}(\bar{v}) = \sum_{i=1}^{n} \sum_{S \subseteq N \setminus i} \frac{s!(n-s-1)!}{n!} (v_L(S \cup i) - v_L(S))$$

and

$$\sum_{i=1}^{n} \phi_{Ri}^{\mathrm{SH}}(\overline{v}) = \sum_{i=1}^{n} \sum_{S \subseteq N \setminus i} \frac{s!(n-s-1)!}{n!} (v_R(S \cup i) - v_R(S)).$$

Thus, in the same way to that of the Shapley value [4, 23], and combining with Definition 1.1, it is straightforwardly derived from the above two equalities that

$$\sum_{i=1}^{n} \phi_{Li}^{\mathrm{SH}}(\overline{v}) = v_L(N)$$

and

$$\sum_{i=1}^{n} \phi_{Ri}^{\mathrm{SH}}(\overline{v}) = v_R(N),$$

i.e.,

$$\sum_{i=1}^{n} \overline{\phi}_i^{\mathrm{SH}}(\overline{v}) = \overline{v}(N).$$

Thus, we have completed the proof of Theorem 3.15.

Theorem 3.16 *(Additivity) For any interval-valued cooperative games* $\overline{v} \in \overline{G}^n$ *and* $\overline{\nu} \in \overline{G}^n$, *if they satisfy Eq. (3.17), then* $\overline{\phi}_i^{\mathrm{SH}}(\overline{v} + \overline{\nu}) = \overline{\phi}_i^{\mathrm{SH}}(\overline{v}) + \overline{\phi}_i^{\mathrm{SH}}(\overline{\nu})$ $(i = 1, 2, \ldots, n)$, *i.e.,* $\overline{\boldsymbol{\Phi}}^{\mathrm{SH}}(\overline{v} + \overline{\nu}) = \overline{\boldsymbol{\Phi}}^{\mathrm{SH}}(\overline{v}) + \overline{\boldsymbol{\Phi}}^{\mathrm{SH}}(\overline{\nu})$.

Proof According to Eq. (3.20) and Definition 1.1, we have

$$\begin{aligned}
\overline{\phi}_i^{\mathrm{SH}}(\overline{v}+\overline{\nu}) = \Bigg[& \sum_{S \subseteq N \setminus i} \frac{s!(n-s-1)!}{n!} [(v_L(S \cup i) + \nu_L(S \cup i)) - (v_L(S) + \nu_L(S))], \\
& \sum_{S \subseteq N \setminus i} \frac{s!(n-s-1)!}{n!} [(v_R(S \cup i) + \nu_R(S \cup i)) - (v_R(S) + \nu_R(S))] \Bigg] \\
= & \left[\sum_{S \subseteq N \setminus i} \frac{s!(n-s-1)!}{n!} (v_L(S \cup i) - v_L(S)), \sum_{S \subseteq N \setminus i} \frac{s!(n-s-1)!}{n!} (v_R(S \cup i) - v_R(S)) \right] \\
& + \left[\sum_{S \subseteq N \setminus i} \frac{s!(n-s-1)!}{n!} (\nu_L(S \cup i) - \nu_L(S)), \sum_{S \subseteq N \setminus i} \frac{s!(n-s-1)!}{n!} (\nu_R(S \cup i) - \nu_R(S)) \right] \\
= & \overline{\phi}_i^{\mathrm{SH}}(\overline{v}) + \overline{\phi}_i^{\mathrm{SH}}(\overline{\nu}),
\end{aligned}$$

i.e.,

$$\overline{\phi}_i^{\mathrm{SH}}(\overline{\upsilon}+\overline{\nu}) = \overline{\phi}_i^{\mathrm{SH}}(\overline{\upsilon}) + \overline{\phi}_i^{\mathrm{SH}}(\overline{\nu}) \quad (i = 1, 2, \ldots, n).$$

Thus, we obtain

$$\overline{\boldsymbol{\Phi}}^{\mathrm{SH}}(\overline{\upsilon}+\overline{\nu}) = \overline{\boldsymbol{\Phi}}^{\mathrm{SH}}(\overline{\upsilon}) + \overline{\boldsymbol{\Phi}}^{\mathrm{SH}}(\overline{\nu}).$$

Therefore, we have completed the proof of Theorem 3.16.

Theorem 3.17 *(Symmetry) For any interval-valued cooperative game $\overline{\upsilon} \in \overline{G}^n$, if it satisfies Eq. (3.17), and players $i \in N$ and $k \in N$ $(i \neq k)$ are symmetric in the interval-valued cooperative game $\overline{\upsilon}$, then $\overline{\phi}_i^{\mathrm{SH}}(\overline{\upsilon}) = \overline{\phi}_k^{\mathrm{SH}}(\overline{\upsilon})$.*

Proof Due to the assumption that the players $i \in N$ and $k \in N (i \neq k)$ are symmetric in the interval-valued cooperative game $\overline{\upsilon} \in \overline{G}^n$, then according to Eq. (3.20) and Definition 1.3, it is straightforward to prove that the conclusion of Theorem 3.17 is valid.

Theorem 3.18 *(Anonymity) For any interval-valued cooperative game $\overline{\upsilon} \in \overline{G}^n$ and any permutation σ on the set N, if $\overline{\upsilon}$ satisfies Eq. (3.17), then $\overline{\phi}_{\sigma(i)}^{\mathrm{SH}}(\overline{\upsilon}^{\sigma}) = \overline{\phi}_i^{\mathrm{SH}}(\overline{\upsilon})$ $(i = 1, 2, \ldots, n)$. Namely, $\overline{\boldsymbol{\Phi}}^{\mathrm{SH}}(\overline{\upsilon}^{\sigma}) = \boldsymbol{\sigma}^{\#}\left(\overline{\boldsymbol{\Phi}}^{\mathrm{SH}}(\overline{\upsilon})\right)$.*

Proof According to Eq. (3.20), we can easily complete the proof of Theorem 3.18 (omitted).

Theorem 3.19 *(Null player) For any interval-valued cooperative game $\overline{\upsilon} \in \overline{G}^n$, if it satisfies Eq. (3.17), and $i \in N$ is a null player in the interval-valued cooperative game $\overline{\upsilon} \in \overline{G}^n$, then $\overline{\phi}_i^{\mathrm{SH}}(\overline{\upsilon}) = 0$.*

Proof According to Eq. (3.20), and combining with Definition 1.4 of the null player i: $\overline{\upsilon}(S \cup i) = \overline{\upsilon}(S)$ for any coalition $S \subseteq N \backslash i$, i.e.,

$$\upsilon_L(S \cup i) = \upsilon_L(S)$$

and

$$\upsilon_R(S \cup i) = \upsilon_R(S),$$

we directly have

$$\begin{aligned}\overline{\phi}_i^{\mathrm{SH}}(\overline{\upsilon}) &= \left[\sum_{S \subseteq N \backslash i} \frac{s!(n-s-1)!}{n!}(\upsilon_L(S \cup i) - \upsilon_L(S)), \sum_{S \subseteq N \backslash i} \frac{s!(n-s-1)!}{n!}(\upsilon_R(S \cup i) - \upsilon_R(S))\right] \\ &= [0, 0] \\ &= 0.\end{aligned}$$

Namely, $\overline{\phi}_i^{\mathrm{SH}}(\overline{\upsilon}) = 0$. Hereby, we have completed the proof of Theorem 3.19.

Theorem 3.20 *(Dummy player) For any interval-valued cooperative game* $\overline{v} \in \overline{G}^n$, *if it satisfies Eq. (3.17), and* $i \in N$ *is a dummy player in the interval-valued cooperative game* $\overline{v} \in \overline{G}^n$, *then* $\overline{\phi}_i^{SH}(\overline{v}) = \overline{v}(i)$.

Proof According to Eq. (3.20), and combining with Definition 1.5 of the dummy player i:

$$\overline{v}(S \cup i) = \overline{v}(S) + \overline{v}(i)$$

for any coalition $S \subseteq N \backslash i$, i.e.,

$$v_L(S \cup i) = v_L(S) + v_L(i)$$

and

$$v_R(S \cup i) = v_R(S) + v_R(i),$$

we can easily have

$$\begin{aligned}
\overline{\phi}_i^{SH}(\overline{v}) &= \left[\sum_{S \subseteq N \backslash i} \frac{s!(n-s-1)!}{n!}(v_L(S \cup i) - v_L(S)), \sum_{S \subseteq N \backslash i} \frac{s!(n-s-1)!}{n!}(v_R(S \cup i) - v_R(S))\right] \\
&= \left[\sum_{S \subseteq N \backslash i} \frac{s!(n-s-1)!}{n!} v_L(i), \sum_{S \subseteq N \backslash i} \frac{s!(n-s-1)!}{n!} v_R(i)\right] \\
&= \left[v_L(i) \sum_{k=0}^{n-1} \sum_{s=k} \frac{s!(n-s-1)!}{n!}, v_R(i) \sum_{k=0}^{n-1} \sum_{s=k} \frac{s!(n-s-1)!}{n!}\right] \\
&= \left[v_L(i) \sum_{k=0}^{n-1} \frac{k!(n-k-1)!}{n!} C_{n-1}^k, v_R(i) \sum_{k=0}^{n-1} \frac{k!(n-k-1)!}{n!} C_{n-1}^k\right] \\
&= \left[v_L(i) \sum_{k=0}^{n-1} \frac{k!(n-k-1)!}{n!} \times \frac{(n-1)!}{k!(n-k-1)!}, v_R(i) \sum_{k=0}^{n-1} \frac{k!(n-k-1)!}{n!} \times \frac{(n-1)!}{k!(n-k-1)!}\right] \\
&= \left[v_L(i) \sum_{k=0}^{n-1} \frac{1}{n}, v_R(i) \sum_{k=0}^{n-1} \frac{1}{n}\right] \\
&= [v_L(i), v_R(i)] \\
&= \overline{v}(i),
\end{aligned}$$

i.e.,

$$\overline{\phi}_i^{SH}(\overline{v}) = \overline{v}(i).$$

Hereby, we have completed the proof of Theorem 3.20.

Theorem 3.21 *(Invariance) For any interval-valued cooperative game* $\overline{v} \in \overline{G}^n$ *and its associated interval-valued cooperative game* $\overline{\nu} \in \overline{G}^n$ *given by Eq. (3.15), if they satisfy Eq. (3.17), then* $\overline{\phi}_i^{\mathrm{SH}}(\overline{\nu}) = a\overline{\phi}_i^{\mathrm{SH}}(\overline{v}) + \overline{d}_i \quad (i = 1, 2, \ldots, n)$, *i.e.,* $\overline{\boldsymbol{\Phi}}^{\mathrm{SH}}(\overline{\nu}) = a\overline{\boldsymbol{\Phi}}^{\mathrm{SH}}(\overline{v}) + \overline{\boldsymbol{d}}$.

Proof According to Eqs. (3.20) and (3.15) and Definition 1.1, we have

$$\overline{\phi}_i^{\mathrm{SH}}(\overline{\nu}) = \left[\sum_{S \subseteq N \setminus i} \frac{s!(n-s-1)!}{n!}\left[\left(av_L(S \cup i) + \sum_{j \in S \cup i} d_{Lj}\right) - \left(av_L(S) + \sum_{j \in S} d_{Lj}\right)\right],\right.$$

$$\left.\sum_{S \subseteq N \setminus i} \frac{s!(n-s-1)!}{n!}\left[\left(av_R(S \cup i) + \sum_{j \in S \cup i} d_{Rj}\right) - \left(av_R(S) + \sum_{j \in S} d_{Rj}\right)\right]\right]$$

$$= a\left[\sum_{S \subseteq N \setminus i} \frac{s!(n-s-1)!}{n!}(v_L(S \cup i) - v_L(S)), \sum_{S \subseteq N \setminus i} \frac{s!(n-s-1)!}{n!}(v_R(S \cup i) - v_R(S))\right]$$

$$+ \left[d_{Li} \sum_{S \subseteq N \setminus i} \frac{s!(n-s-1)!}{n!}, d_{Ri} \sum_{S \subseteq N \setminus i} \frac{s!(n-s-1)!}{n!}\right]$$

$$= a\overline{\phi}_i^{\mathrm{SH}}(\overline{v}) + \left[d_{Li} \sum_{k=0}^{n-1} \frac{k!(n-k-1)!}{n!} \times \frac{(n-1)!}{k!(n-k-1)!},\right.$$

$$\left. d_{Ri} \sum_{k=0}^{n-1} \frac{k!(n-k-1)!}{n!} \times \frac{(n-1)!}{k!(n-k-1)!}\right]$$

$$= a\overline{\phi}_i^{\mathrm{SH}}(\overline{v}) + \overline{d}_i,$$

i.e.,

$$\overline{\phi}_i^{\mathrm{SH}}(\overline{\nu}) = a\overline{\phi}_i^{\mathrm{SH}}(\overline{v}) + \overline{d}_i \quad (i = 1, 2, \ldots, n).$$

Hereby, we obtain

$$\overline{\boldsymbol{\Phi}}^{\mathrm{SH}}(\overline{\nu}) = a\overline{\boldsymbol{\Phi}}^{\mathrm{SH}}(\overline{v}) + \overline{\boldsymbol{d}}.$$

Thus, we have completed the proof of Theorem 3.21.

Furthermore, interval-valued Shapley values of interval-valued cooperative games do not always satisfy the individual rationality. A specific illustrated example may be referred to Example 3.6 as above.

3.3.2 Interval-Valued Egalitarian Shapley Values of Interval-Valued Cooperative Games and Properties

For any cooperative game $v \in G^n$, combining the equal division value (or egalitarian value) with the Shapley value, Joosten [24] introduced the egalitarian Shapley value:

$$\boldsymbol{\Phi}^{\mathrm{ESH}\zeta}(v) = \left(\phi_1^{\mathrm{ESH}\zeta}(v), \phi_2^{\mathrm{ESH}\zeta}(v), \ldots, \phi_n^{\mathrm{ESH}\zeta}(v)\right)^{\mathrm{T}},$$

whose components are given as follows:

$$\phi_i^{\mathrm{ESH}\zeta}(v) = (1-\zeta)\rho_i^{\mathrm{ED}}(v) + \zeta\phi_i^{\mathrm{SH}}(v) \quad (i = 1, 2, \ldots, n), \tag{3.27}$$

where the parameter $\zeta \in [0, 1]$ may be chosen by the players according to need in real situations, $\phi_i^{\mathrm{SH}}(v)$ and $\rho_i^{\mathrm{ED}}(v)$ $(i = 1, 2, \ldots, n)$ are given by Eqs. (1.3) and (3.1), respectively.

Analogously, for any interval-valued cooperative game $\overline{v} \in \overline{G}^n$, if it satisfies Eq. (3.17), then we can define the interval-valued egalitarian Shapley value $\overline{\boldsymbol{\Phi}}^{\mathrm{ESH}\zeta}(\overline{v})$ as follows:

$$\overline{\boldsymbol{\Phi}}^{\mathrm{ESH}\zeta}(\overline{v}) = (1-\zeta)\overline{\boldsymbol{\rho}}^{\mathrm{ED}}(\overline{v}) + \zeta\overline{\boldsymbol{\Phi}}^{\mathrm{SH}}(\overline{v}), \tag{3.28}$$

where $\overline{\boldsymbol{\rho}}^{\mathrm{ED}}(\overline{v})$ and $\overline{\boldsymbol{\Phi}}^{\mathrm{SH}}(\overline{v})$ are given by Eqs. (3.3) and (3.20) (or Eqs. (3.18) and (3.19)), respectively.

Obviously, if $\zeta = 0$, then the interval-valued egalitarian Shapley value $\overline{\boldsymbol{\Phi}}^{\mathrm{ESH}\zeta}(\overline{v})$ is reduced to the interval-valued equal division value $\overline{\boldsymbol{\rho}}^{\mathrm{ED}}(\overline{v})$, i.e., $\overline{\boldsymbol{\Phi}}^{\mathrm{ESH0}}(\overline{v}) = \overline{\boldsymbol{\rho}}^{\mathrm{ED}}(\overline{v})$; if $\zeta = 1$, then the interval-valued egalitarian Shapley value $\overline{\boldsymbol{\Phi}}^{\mathrm{ESH}\zeta}(\overline{v})$ is reduced to the interval-valued Shapley value $\overline{\boldsymbol{\Phi}}^{\mathrm{SH}}(\overline{v})$, i.e., $\overline{\boldsymbol{\Phi}}^{\mathrm{ESH1}}(\overline{v}) = \overline{\boldsymbol{\Phi}}^{\mathrm{SH}}(\overline{v})$.

Example 3.8 Let us compute the interval-valued egalitarian Shapley value of the interval-valued cooperative game $\overline{v}^0 \in \overline{G}^3$ given in Example 3.6.

It is easy to see from Example 3.6 that $\overline{v}^0(N') = [2, 7]$ and the interval-valued Shapley value of the interval-valued cooperative game $\overline{v}^0 \in \overline{G}^3$ is equal to

$$\overline{\boldsymbol{\Phi}}^{\mathrm{SH}}(\overline{v}^0) = \left(\left[0, \frac{7}{6}\right], \left[1, \frac{8}{3}\right], \left[1, \frac{19}{6}\right]\right)^{\mathrm{T}},$$

where $N' = \{1, 2, 3\}$. Thus, according to Eq. (3.28), the interval-valued egalitarian Shapley value of the interval-valued cooperative game $\overline{v}^0 \in \overline{G}^3$ can be obtained as follows:

$$\overline{\Phi}^{\mathrm{ESH}\zeta}(\overline{v}^0) = (1-\zeta)\left(\frac{\overline{v}^0(N')}{3}, \frac{\overline{v}^0(N')}{3}, \frac{\overline{v}^0(N')}{3}\right)^{\mathrm{T}} + \zeta\overline{\Phi}^{\mathrm{SH}}(\overline{v}^0)$$
$$= (1-\zeta)\left(\left[\frac{2}{3}, \frac{7}{3}\right], \left[\frac{2}{3}, \frac{7}{3}\right], \left[\frac{2}{3}, \frac{7}{3}\right]\right)^{\mathrm{T}} + \zeta\left(\left[0, \frac{7}{6}\right], \left[1, \frac{8}{3}\right], \left[1, \frac{19}{6}\right]\right)^{\mathrm{T}}$$
$$= \left(\left[\frac{2(1-\zeta)}{3}, \frac{14-7\zeta}{6}\right], \left[\frac{2+\zeta}{3}, \frac{7+\zeta}{3}\right], \left[\frac{2+\zeta}{3}, \frac{14+5\zeta}{6}\right]\right)^{\mathrm{T}},$$

where $\zeta \in [0, 1]$.

Similarly, we can easily obtain some useful and important properties of interval-valued egalitarian Shapley values of interval-valued cooperative games through considering the properties of interval-valued Shapley values and interval-valued equal division values.

Theorem 3.22 *(Existence and Uniqueness) For an arbitrary interval-valued cooperative game $\overline{v} \in \overline{G}^n$ and a given parameter $\zeta \in [0, 1]$, if $\overline{v}$ satisfies Eq. (3.17), then there always exists a unique interval-valued egalitarian Shapley value $\overline{\Phi}^{\mathrm{ESH}\zeta}(\overline{v})$, which is determined by Eq. (3.28).*

Proof According to Eq. (3.28), and combining with Theorems 3.1 and 3.14, we have completed the proof of Theorem 3.22.

Theorem 3.23 *(Efficiency) For any interval-valued cooperative game $\overline{v} \in \overline{G}^n$, if it satisfies Eq. (3.17), then its interval-valued egalitarian Shapley value $\overline{\Phi}^{\mathrm{ESH}\zeta}(\overline{v})$ satisfies the efficiency, i.e., $\sum_{i=1}^{n} \overline{\phi}_i^{\mathrm{ESH}\zeta}(\overline{v}) = \overline{v}(N)$.*

Proof According to Eq. (3.28), and combining with Theorems 3.2 and 3.15 and Definition 1.1, we have

$$\sum_{i=1}^{n} \overline{\phi}_i^{\mathrm{ESH}\zeta}(\overline{v}) = \sum_{i=1}^{n}\left[(1-\zeta)\overline{\rho}_i^{\mathrm{ED}}(\overline{v}) + \zeta\overline{\phi}_i^{\mathrm{SH}}(\overline{v})\right]$$
$$= (1-\zeta)\sum_{i=1}^{n}\overline{\rho}_i^{\mathrm{ED}}(\overline{v}) + \zeta\sum_{i=1}^{n}\overline{\phi}_i^{\mathrm{SH}}(\overline{v})$$
$$= (1-\zeta)\overline{v}(N) + \zeta\overline{v}(N)$$
$$= \overline{v}(N),$$

i.e.,

$$\sum_{i=1}^{n} \overline{\phi}_i^{\mathrm{ESH}\zeta}(\overline{v}) = \overline{v}(N).$$

Thus, we have completed the proof of Theorem 3.23.

Theorem 3.24 *(Additivity) For any interval-valued cooperative games* $\overline{\upsilon} \in \overline{G}^n$ *and* $\overline{\nu} \in \overline{G}^n$, *if they satisfy Eq. (3.17), then* $\overline{\phi}_i^{\text{ESH}\zeta}(\overline{\upsilon} + \overline{\nu}) = \overline{\phi}_i^{\text{ESH}\zeta}(\overline{\upsilon}) + \overline{\phi}_i^{\text{ESH}\zeta}(\overline{\nu})$ $(i = 1, 2, \ldots, n)$, *i.e.,* $\overline{\boldsymbol{\Phi}}^{\text{ESH}\zeta}(\overline{\upsilon} + \overline{\nu}) = \overline{\boldsymbol{\Phi}}^{\text{ESH}\zeta}(\overline{\upsilon}) + \overline{\boldsymbol{\Phi}}^{\text{ESH}\zeta}(\overline{\nu})$.

Proof According to Eq. (3.28), and combining with Theorems 3.3 and 3.16 and Definition 1.1, we have

$$\begin{aligned}
\overline{\boldsymbol{\Phi}}^{\text{ESH}\zeta}(\overline{\upsilon} + \overline{\nu}) &= (1 - \zeta)\overline{\boldsymbol{\rho}}^{\text{ED}}(\overline{\upsilon} + \overline{\nu}) + \zeta\overline{\boldsymbol{\Phi}}^{\text{SH}}(\overline{\upsilon} + \overline{\nu}) \\
&= \left[(1 - \zeta)\overline{\boldsymbol{\rho}}^{\text{ED}}(\overline{\upsilon}) + (1 - \zeta)\overline{\boldsymbol{\rho}}^{\text{ED}}(\overline{\nu})\right] + \left(\zeta\overline{\boldsymbol{\Phi}}^{\text{SH}}(\overline{\upsilon}) + \zeta\overline{\boldsymbol{\Phi}}^{\text{SH}}(\overline{\nu})\right) \\
&= \left[(1 - \zeta)\overline{\boldsymbol{\rho}}^{\text{ED}}(\overline{\upsilon}) + \zeta\overline{\boldsymbol{\Phi}}^{\text{SH}}(\overline{\upsilon})\right] + \left[(1 - \zeta)\overline{\boldsymbol{\rho}}^{\text{ED}}(\overline{\nu}) + \zeta\overline{\boldsymbol{\Phi}}^{\text{SH}}(\overline{\nu})\right] \\
&= \overline{\boldsymbol{\Phi}}^{\text{ESH}\zeta}(\overline{\upsilon}) + \overline{\boldsymbol{\Phi}}^{\text{ESH}\zeta}(\overline{\nu}),
\end{aligned}$$

i.e.,

$$\overline{\boldsymbol{\Phi}}^{\text{ESH}\zeta}(\overline{\upsilon} + \overline{\nu}) = \overline{\boldsymbol{\Phi}}^{\text{ESH}\zeta}(\overline{\upsilon}) + \overline{\boldsymbol{\Phi}}^{\text{ESH}\zeta}(\overline{\nu}).$$

Therefore, we have completed the proof of Theorem 3.24.

Theorem 3.25 (Symmetry) For any interval-valued cooperative game $\overline{\upsilon} \in \overline{G}^n$, if it satisfies Eq. (3.17), and players $i \in N$ and $k \in N$ $(i \neq k)$ are symmetric in $\overline{\upsilon} \in \overline{G}^n$, then $\overline{\phi}_i^{\text{ESH}\zeta}(\overline{\upsilon}) = \overline{\phi}_k^{\text{ESH}\zeta}(\overline{\upsilon})$.

Proof According to Eq. (3.28), and combining with Theorems 3.4 and 3.17 and Definitions 1.1 and 1.3, we have

$$\begin{aligned}
\overline{\phi}_i^{\text{ESH}\zeta}(\overline{\upsilon}) &= (1 - \zeta)\overline{\rho}_i^{\text{ED}}(\overline{\upsilon}) + \zeta\overline{\phi}_i^{\text{SH}}(\overline{\upsilon}) \\
&= (1 - \zeta)\overline{\rho}_k^{\text{ED}}(\overline{\upsilon}) + \zeta\overline{\phi}_k^{\text{SH}}(\overline{\upsilon}) \\
&= \overline{\phi}_k^{\text{ESH}\zeta}(\overline{\upsilon}),
\end{aligned}$$

i.e.,

$$\overline{\phi}_i^{\text{ESH}\zeta}(\overline{\upsilon}) = \overline{\phi}_k^{\text{ESH}\zeta}(\overline{\upsilon}).$$

Thus, we have completed the proof of Theorem 3.25.

Theorem 3.26 (Anonymity) For any interval-valued cooperative game $\overline{\upsilon} \in \overline{G}^n$ and any permutation σ on the set N, if $\overline{\upsilon}$ satisfies Eq. (3.17), then $\overline{\phi}_{\sigma(i)}^{\text{ESH}\zeta}(\overline{\upsilon}^\sigma) = \overline{\phi}_i^{\text{ESH}\zeta}(\overline{\upsilon})$ $(i = 1, 2, \ldots, n)$. Namely, $\overline{\boldsymbol{\Phi}}^{\text{ESH}\zeta}(\overline{\upsilon}^\sigma) = \boldsymbol{\sigma}^{\#}\left(\overline{\boldsymbol{\Phi}}^{\text{ESH}\zeta}(\overline{\upsilon})\right)$.

Proof According to Eq. (3.28), and combining with Theorems 3.5 and 3.18, we have

$$\begin{aligned}\overline{\boldsymbol{\Phi}}^{\mathrm{ESH}\zeta}(\overline{v}^{\sigma}) &= (1-\zeta)\overline{\boldsymbol{\rho}}^{\mathrm{ED}}(\overline{v}^{\sigma}) + \zeta\overline{\boldsymbol{\Phi}}^{\mathrm{SH}}(\overline{v}^{\sigma}) \\ &= (1-\zeta)\boldsymbol{\sigma}^{\#}\left(\overline{\boldsymbol{\rho}}^{\mathrm{ED}}(\overline{v})\right) + \zeta\boldsymbol{\sigma}^{\#}\left(\overline{\boldsymbol{\Phi}}^{\mathrm{SH}}(\overline{v})\right) \\ &= \boldsymbol{\sigma}^{\#}\left((1-\zeta)\overline{\boldsymbol{\rho}}^{\mathrm{ED}}(\overline{v}) + \zeta\overline{\boldsymbol{\Phi}}^{\mathrm{SH}}(\overline{v})\right) \\ &= \boldsymbol{\sigma}^{\#}\left(\overline{\boldsymbol{\Phi}}^{\mathrm{ESH}\zeta}(\overline{v})\right),\end{aligned}$$

i.e.,

$$\overline{\boldsymbol{\Phi}}^{\mathrm{ESH}\zeta}(\overline{v}^{\sigma}) = \boldsymbol{\sigma}^{\#}\left(\overline{\boldsymbol{\Phi}}^{\mathrm{ESH}\zeta}(\overline{v})\right).$$

Thus, we have completed the proof of Theorem 3.26.

Generally, interval-valued egalitarian Shapley values of interval-valued cooperative games do not satisfy the dummy player property, the null player property, and the invariance although interval-valued Shapley values do. Moreover, interval-valued egalitarian Shapley values do not always satisfy the individual rationality.

3.3.3 Interval-Valued Discounted Shapley Values of Interval-Valued Cooperative Games and Properties

In crisp (or classic) cooperative games, the Shapley value was extended to the discounted Shapley value [24, 25]. Specifically, for any cooperative game $v \in G^n$, its discounted Shapley value is defined as a payoff vector

$$\boldsymbol{\Phi}^{\mathrm{DSH}\delta}(v) = \left(\phi_1^{\mathrm{DSH}\delta}(v), \phi_2^{\mathrm{DSH}\delta}(v), \ldots, \phi_n^{\mathrm{DSH}\delta}(v)\right)^{\mathrm{T}},$$

whose components are given as follows:

$$\phi_i^{\mathrm{DSH}\delta}(v) = \sum_{S \subseteq N \setminus i} \frac{s!(n-s-1)!}{n!} \delta^{n-s-1}(v(S \cup i) - \delta v(S)) \quad (i = 1, 2, \ldots, n), \tag{3.29}$$

respectively, where $\delta \in [0,1]$ is a discounted factor [26]. As customary, we stipulate: $0^0 = 1$.

In the sequent, let us continue to consider the interval-valued cooperative game $\overline{v} \in \overline{G}^n$ given in Sect. 1.3.2.

Analogously, we can construct an associated cooperative game $v(\alpha) \in G^n$, where the set of players is $N = \{1, 2, \ldots, n\}$ and the characteristic function $v(\alpha)$ of coalitions of players is given by Eq. (3.4).

According to Eq. (3.29), we can easily obtain the discounted Shapley value $\boldsymbol{\Phi}^{\mathrm{DSH}\delta}(v(\alpha)) = (\phi_1^{\mathrm{DSH}\delta}(v(\alpha)), \phi_2^{\mathrm{DSH}\delta}(v(\alpha)), \ldots, \phi_n^{\mathrm{DSH}\delta}(v(\alpha)))^{\mathrm{T}}$ of the cooperative game $v(\alpha) \in G^n$, where

$$\phi_i^{\mathrm{DSH}\delta}(v(\alpha)) = \sum_{S \subseteq N \setminus i} \frac{s!(n-s-1)!}{n!} \delta^{n-s-1} (v(\alpha)(S \cup i) - \delta v(\alpha)(S)) \quad (i = 1, 2, \ldots, n),$$

which can further be rewritten as follows:

$$\phi_i^{\mathrm{DSH}\delta}(v(\alpha)) = \sum_{S \subseteq N \setminus i} \frac{s!(n-s-1)!}{n!} \delta^{n-s-1} \{[(1-\alpha) v_L(S \cup i) + \alpha v_R(S \cup i)] \\ - \delta[(1-\alpha) v_L(S) + \alpha v_R(S)]\} \quad (i = 1, 2, \ldots, n), \tag{3.30}$$

where $\alpha \in [0, 1]$.

It is clear that the discounted Shapley value $\boldsymbol{\Phi}^{\mathrm{DSH}\delta}(v)$ is a continuous function of the parameter $\alpha \in [0, 1]$.

Theorem 3.27 *For any interval-valued cooperative game $\overline{v} \in \overline{G}^n$ and a given parameter $\delta \in [0, 1]$, if the following system of inequalities*

$$v_R(S \cup i) - v_L(S \cup i) \geq \delta(v_R(S) - v_L(S)) \quad (i = 1, 2, \ldots, n; S \subseteq N \setminus i) \tag{3.31}$$

is satisfied, then the discounted Shapley value $\boldsymbol{\Phi}^{\mathrm{DSH}\delta}(v)$ of the cooperative game $v(\alpha) \in G^n$ is a monotonic and non-decreasing function of the parameter $\alpha \in [0, 1]$.

Proof For any $\alpha \in [0, 1]$ and $\alpha' \in [0, 1]$, according to Eq. (3.30), we have

$$\begin{aligned}\phi_i^{\mathrm{DSH}\delta}(v(\alpha)) - \phi_i^{\mathrm{DSH}\delta}(v(\alpha')) &= \sum_{S \subseteq N \setminus i} \left\{ \frac{s!(n-s-1)!}{n!} \delta^{n-s-1} \right. \\ &\quad \left. \times \left[(v(\alpha)(S \cup i) - v(\alpha')(S \cup i)) - \delta(v(\alpha)(S) - v(\alpha')(S)) \right] \right\} \\ &= (\alpha - \alpha') \sum_{S \subseteq N \setminus i} \left\{ \frac{s!(n-s-1)!}{n!} \delta^{n-s-1} \right. \\ &\quad \left. \times [(v_R(S \cup i) - v_L(S \cup i)) - \delta(v_R(S) - v_L(S))] \right\},\end{aligned}$$

where $i = 1, 2, \ldots, n$.

If $\alpha \geq \alpha'$, then combining with the assumption, i.e., Eq. (3.31), we have

$$\phi_i^{\mathrm{DSH}\delta}(v(\alpha)) - \phi_i^{\mathrm{DSH}\delta}\left(v\left(\alpha'\right)\right) \geq 0 \quad (i = 1, 2, \ldots, n),$$

i.e.,

$$\phi_i^{\mathrm{DSH}\delta}(v(\alpha)) \geq \phi_i^{\mathrm{DSH}\delta}\left(v\left(\alpha'\right)\right) \quad (i = 1, 2, \ldots, n),$$

which mean that the discounted Shapley value $\boldsymbol{\Phi}^{\mathrm{DSH}\delta}(v)$ is a monotonic and non-decreasing function of the parameter $\alpha \in [0, 1]$. Thus, we have completed the proof of Theorem 3.27.

Accordingly, for any interval-valued cooperative game $\overline{v} \in \overline{G}^n$, if it satisfies Eq. (3.31), then it is directly derived from Theorem 3.27 and Eq. (3.30) that the lower and upper bounds of the components (intervals) $\overline{\phi}_i^{\mathrm{DSH}\delta}(\overline{v})$ $(i = 1, 2, \ldots, n)$ of the interval-valued discounted Shapley value $\overline{\boldsymbol{\Phi}}^{\mathrm{DSH}\delta}(\overline{v}) = \left(\overline{\phi}_1^{\mathrm{DSH}\delta}(\overline{v}), \overline{\phi}_2^{\mathrm{DSH}\delta}(\overline{v}), \ldots, \overline{\phi}_n^{\mathrm{DSH}\delta}(\overline{v})\right)^{\mathrm{T}}$ are given as follows:

$$\begin{aligned}\phi_{Li}^{\mathrm{DSH}\delta}(\overline{v}) &= \phi_i^{\mathrm{DSH}\delta}(v(0)) \\ &= \sum_{S \subseteq N \setminus i} \frac{s!(n-s-1)!}{n!} \delta^{n-s-1}(v_L(S \cup i) - \delta v_L(S)) \quad (i = 1, 2, \ldots, n)\end{aligned}$$

and

$$\begin{aligned}\phi_{Ri}^{\mathrm{DSH}\delta}(\overline{v}) &= \phi_i^{\mathrm{DSH}\delta}(v(1)) \\ &= \sum_{S \subseteq N \setminus i} \frac{s!(n-s-1)!}{n!} \delta^{n-s-1}(v_R(S \cup i) - \delta v_R(S)) \quad (i = 1, 2, \ldots, n),\end{aligned}$$

respectively, i.e.,

$$\phi_{Li}^{\mathrm{DSH}\delta}(\overline{v}) = \sum_{S \subseteq N \setminus i} \frac{s!(n-s-1)!}{n!} \delta^{n-s-1}(v_L(S \cup i) - \delta v_L(S)) \quad (i = 1, 2, \ldots, n) \tag{3.32}$$

and

$$\phi_{Ri}^{\mathrm{DSH}\delta}(\overline{v}) = \sum_{S \subseteq N \setminus i} \frac{s!(n-s-1)!}{n!} \delta^{n-s-1}(v_R(S \cup i) - \delta v_R(S)) \quad (i = 1, 2, \ldots, n). \tag{3.33}$$

Thus, the interval-valued discounted Shapley values $\overline{\phi}_i^{\mathrm{DSH}\delta}(\overline{v}) = [\phi_{Li}^{\mathrm{DSH}\delta}(\overline{v}), \phi_{Ri}^{\mathrm{DSH}\delta}(\overline{v})]$ of the players i $(i = 1, 2, \ldots, n)$ in the interval-valued cooperative game $\overline{v} \in \overline{G}^n$ are directly and explicitly expressed as follows:

$$\overline{\phi}_i^{\mathrm{DSH}\delta}(\overline{v}) = \left[\sum_{S \subseteq N \setminus i} \frac{s!(n-s-1)!}{n!} \delta^{n-s-1}(v_L(S \cup i) - \delta v_L(S)), \right.$$

$$\left. \sum_{S \subseteq N \setminus i} \frac{s!(n-s-1)!}{n!} \delta^{n-s-1}(v_R(S \cup i) - \delta v_R(S)) \right] \quad (i = 1, 2, \ldots, n). \tag{3.34}$$

Clearly, if $\delta = 1$, then the interval-valued discounted Shapley value $\overline{\boldsymbol{\Phi}}^{\mathrm{DSH}\delta}(\overline{v})$ is reduced to the interval-valued Shapley value $\overline{\boldsymbol{\Phi}}^{\mathrm{SH}}(\overline{v})$, i.e., $\overline{\boldsymbol{\Phi}}^{\mathrm{DSH1}}(\overline{v}) = \overline{\boldsymbol{\Phi}}^{\mathrm{SH}}(\overline{v})$; if $\delta = 0$, then the interval-valued discounted Shapley value $\overline{\boldsymbol{\Phi}}^{\mathrm{DSH}\delta}(\overline{v})$ is reduced to the interval-valued equal division value $\overline{\boldsymbol{\rho}}^{\mathrm{ED}}(\overline{v})$, i.e., $\overline{\boldsymbol{\Phi}}^{\mathrm{DSH0}}(\overline{v}) = \overline{\boldsymbol{\rho}}^{\mathrm{ED}}(\overline{v})$.

Equation (3.31) is an important condition which ensures that the interval-valued cooperative game $\overline{v} \in \overline{G}^n$ has the interval-valued discounted Shapley value $\overline{\boldsymbol{\Phi}}^{\mathrm{DSH}\delta}(\overline{v})$ given by Eq. (3.34) (or Eqs. (3.32) and (3.33)). The condition expressed by Eq. (3.31) is weaker than that by Eq. (3.17) due to the fact that the former is always true if the latter is satisfied.

Particularly, for any two-person interval-valued cooperative game $\overline{v} \in \overline{G}^2$, if it satisfies Eq. (3.31), then according to Eqs. (3.32) and (3.33), we can explicitly rewrite the lower and upper bounds of the interval-valued discounted Shapley values $\overline{\phi}_i^{\mathrm{DSH}\delta}(\overline{v})$ of the players i $(i = 1, 2)$ in the interval-valued cooperative game $\overline{v} \in \overline{G}^2$ as follows:

$$\phi_{L1}^{\mathrm{DSH}\delta}(\overline{v}) = \frac{\delta v_L(1) + v_L(1,2) - \delta v_L(2)}{2}, \tag{3.35}$$

$$\phi_{R1}^{\mathrm{DSH}\delta}(\overline{v}) = \frac{\delta v_R(1) + v_R(1,2) - \delta v_R(2)}{2}, \tag{3.36}$$

$$\phi_{L2}^{\mathrm{DSH}\delta}(\overline{v}) = \frac{\delta v_L(2) + v_L(1,2) - \delta v_L(1)}{2}, \tag{3.37}$$

and

$$\phi_{R2}^{\mathrm{DSH}\delta}(\overline{v}) = \frac{\delta v_R(2) + v_R(1,2) - \delta v_R(1)}{2}. \tag{3.38}$$

Therefore, we can simply rewrite the interval-valued discounted Shapley values $\overline{\phi}_i^{\mathrm{DSH}\delta}(\overline{v})$ of the players i $(i = 1, 2)$ in the interval-valued cooperative game $\overline{v} \in \overline{G}^2$ as follows:

$$\overline{\phi}_1^{\mathrm{DSH}\delta}(\overline{v}) = \left[\frac{\delta v_L(1) + v_L(1,2) - \delta v_L(2)}{2}, \frac{\delta v_R(1) + v_R(1,2) - \delta v_R(2)}{2}\right]$$

and

$$\overline{\phi}_2^{\mathrm{DSH}\delta}(\overline{v}) = \left[\frac{\delta v_L(2) + v_L(1,2) - \delta v_L(1)}{2}, \frac{\delta v_R(2) + v_R(1,2) - \delta v_R(1)}{2}\right].$$

Alternatively, the above two equalities can be rewritten as follows:

$$\overline{\phi}_1^{\mathrm{DSH}\delta}(\overline{v}) = \left[\delta v_L(1) + \frac{v_L(1,2) - \delta(v_L(1) + v_L(2))}{2}, \delta v_R(1) + \frac{v_R(1,2) - \delta(v_R(1) + v_R(2))}{2}\right]$$

and

$$\overline{\phi}_2^{\mathrm{DSH}\delta}(\overline{v}) = \left[\delta v_L(2) + \frac{v_L(1,2) - \delta(v_L(1) + v_L(2))}{2}, \delta v_R(2) + \frac{v_R(1,2) - \delta(v_R(1) + v_R(2))}{2}\right].$$

Hereby, inspired by the idea of the interval-valued equal surplus division value given by Eq. (3.14), if an interval-valued cooperative game $\overline{v} \in \overline{G}^n$ satisfies the following system of inequalities

$$v_R(N) - v_L(N) \geq \delta \sum_{j=1}^{n} \left[(v_R(j) - v_L(j)) - (v_R(i) - v_L(i))\right] \quad (i = 1, 2, \ldots, n), \tag{3.39}$$

namely,

$$v_R(N) - v_L(N) \geq -n\delta(v_R(i) - v_L(i)) + \delta \sum_{j=1}^{n} (v_R(j) - v_L(j)) \quad (i = 1, 2, \ldots, n), \tag{3.40}$$

then we can similarly define its interval-valued discounted equal surplus division value as follows:

$$\overline{\rho}^{\mathrm{DESD}\delta}(\overline{v}) = \left(\overline{\rho}_1^{\mathrm{DESD}\delta}(\overline{v}), \overline{\rho}_2^{\mathrm{DESD}\delta}(\overline{v}), \ldots, \overline{\rho}_n^{\mathrm{DESD}\delta}(\overline{v})\right)^{\mathrm{T}},$$

whose components are given as follows:

$$\overline{\rho}_i^{\mathrm{DESD}\delta}(\overline{v}) = \left[\delta v_L(i) + \frac{v_L(N) - \delta \sum_{j=1}^{n} v_L(j)}{n}, \delta v_R(i) + \frac{v_R(N) - \delta \sum_{j=1}^{n} v_R(j)}{n}\right] \quad (i = 1, 2, \ldots, n). \tag{3.41}$$

Clearly, if $\delta = 1$, then the interval-valued discounted equal surplus division value $\bar{\rho}^{\text{DESD}\delta}(\bar{v})$ is reduced to the interval-valued equal surplus division value $\bar{\rho}^{\text{ESD}}(\bar{v})$, i.e., $\bar{\rho}^{\text{DESD1}}(\bar{v}) = \bar{\rho}^{\text{ESD}}(\bar{v})$; if $\delta = 0$, then the interval-valued discounted equal surplus division value $\bar{\rho}^{\text{DESD}\delta}(\bar{v})$ is reduced to the interval-valued equal division value $\bar{\rho}^{\text{ED}}(\bar{v})$, i.e., $\bar{\rho}^{\text{DESD0}}(\bar{v}) = \bar{\rho}^{\text{ED}}(\bar{v})$.

Analogously, for a given parameter δ ($\delta \neq 1$ or 0), if an interval-valued cooperative game $\bar{v} \in \overline{G}^n$ satisfies Eq. (3.39) (or Eq. (3.40)), then we can verify that its interval-valued discounted equal surplus division value $\bar{\rho}^{\text{DESD}\delta}(\bar{v})$ satisfies the efficiency, the symmetry, the additivity, and the anonymity. However, $\bar{\rho}^{\text{DESD}\delta}(\bar{v})$ does not always satisfy the invariance, the individual rationality, the dummy player property, and the null player property.

Example 3.9 Let us compute the interval-valued discounted Shapley value $\boldsymbol{\Phi}^{\text{DSH},0.5}(\bar{v}'')$ and the interval-valued discounted equal surplus division value $\bar{\rho}^{\text{DESD},0.5}(\bar{v}'')$ of the interval-valued cooperative game $\bar{v}'' \in \overline{G}^2$ given in Example 1.2.

From Example 1.2, we have $\bar{v}''(1) = [0.3, 1], \bar{v}''(2) = [2, 5]$, and $\bar{v}''(1, 2) = [4, 6]$, where $N'' = \{1, 2\}$ and $\bar{v}''(\varnothing) = 0$. Therefore, we obtain

$$\bar{v}''_R(1,2) - \bar{v}''_L(1,2) = 6 - 4 = 2 > \delta''\left(\bar{v}''_R(1) - \bar{v}''_L(1)\right) = 0.5 \times (1 - 0.3) = 0.35$$

and

$$\bar{v}''_R(1,2) - \bar{v}''_L(1,2) = 6 - 4 = 2 > \delta''\left(\bar{v}''_R(2) - \bar{v}''_L(2)\right) = 0.5 \times (5 - 2) = 1.5,$$

where $\delta'' = 0.5$. That is to say, the interval-valued cooperative game $\bar{v}'' \in \overline{G}^2$ satisfies Eq. (3.31). Thus, using Eqs. (3.35)–(3.38), we have

$$\phi_{L1}^{\text{DSH},0.5}\left(\bar{v}''\right) = \frac{0.3\delta'' + 4 - 2\delta''}{2} = 1.575,$$

$$\phi_{R1}^{\text{DSH},0.5}\left(\bar{v}''\right) = \frac{\delta'' + 6 - 5\delta''}{2} = 2,$$

$$\phi_{L2}^{\text{DSH},0.5}\left(\bar{v}''\right) = \frac{2\delta'' + 4 - 0.3\delta''}{2} = 2.425,$$

and

$$\phi_{R2}^{\text{DSH},0.5}\left(\bar{v}''\right) = \frac{5\delta'' + 6 - \delta''}{2} = 4.$$

Hence, we obtain the interval-valued discounted Shapley value $\boldsymbol{\Phi}^{\text{DSH},0.5}(\bar{v}'') = ([1.575, 2], [2.425, 4])^{\text{T}}$ of the interval-valued cooperative game $\bar{v}'' \in \overline{G}^2$. However, it is easy to see from Example 3.7 that the interval-valued cooperative game $\overline{v}'' \in \overline{G}^2$ has not the interval-valued Shapley value given by Eq. (3.20) (or Eqs. (3.25) and (3.26)).

Similarly, we can easily compute

$$-2\delta''\left(\overline{v}''_R(1)-\overline{v}''_L(1)\right)+\delta''\left[\left(\overline{v}''_R(1)-\overline{v}''_L(1)\right)+\left(\overline{v}''_R(2)-\overline{v}''_L(2)\right)\right]=1.15$$

and

$$-2\delta''\left(\overline{v}''_R(2)-\overline{v}''_L(2)\right)+\delta''\left[\left(\overline{v}''_R(1)-\overline{v}''_L(1)\right)+\left(\overline{v}''_R(2)-\overline{v}''_L(2)\right)\right]=-1.15.$$

Hence, we have

$$\begin{aligned}\overline{v}''_R(1,2)-\overline{v}''_L(1,2)>-2\delta''&\left(\overline{v}''_R(1)-\overline{v}''_L(1)\right)\\&+\delta''\left[\left(\overline{v}''_R(1)-\overline{v}''_L(1)\right)+\left(\overline{v}''_R(2)-\overline{v}''_L(2)\right)\right]\end{aligned}$$

and

$$\begin{aligned}\overline{v}''_R(1,2)-\overline{v}''_L(1,2)>-2\delta''&\left(\overline{v}''_R(2)-\overline{v}''_L(2)\right)\\&+\delta''\left[\left(\overline{v}''_R(1)-\overline{v}''_L(1)\right)+\left(\overline{v}''_R(2)-\overline{v}''_L(2)\right)\right].\end{aligned}$$

Therefore, the interval-valued cooperative game $\overline{v}'' \in \overline{G}^2$ satisfies Eq. (3.40). Then, using Eq. (3.41), we have

$$\begin{aligned}\overline{\rho}_1^{\mathrm{DESD},0.5}(\overline{v}'')&=\left[0.3\delta''+\frac{4-\delta''(0.3+2)}{2},\delta''+\frac{6-\delta''(1+5)}{2}\right]\\&=\left[2-0.85\delta'',3-2\delta''\right]\\&=[1.575,2]\end{aligned}$$

and

$$\begin{aligned}\overline{\rho}_2^{\mathrm{DESD},0.5}(\overline{v}'')&=\left[2\delta''+\frac{4-\delta''(0.3+2)}{2},5\delta''+\frac{6-\delta''(1+5)}{2}\right]\\&=\left[2+0.85\delta'',3+2\delta''\right]\\&=[2.425,4].\end{aligned}$$

Thereby, we obtain the interval-valued discounted equal surplus division value as follows:

$$\overline{\rho}^{\mathrm{DESD},0.5}\left(\overline{\upsilon}''\right) = ([1.575, 2], [2.425, 4])^{\mathrm{T}}.$$

Obviously, we have

$$\overline{\boldsymbol{\Phi}}^{\mathrm{DSH},0.5}\left(\overline{\upsilon}''\right) = \overline{\rho}^{\mathrm{DESD},0.5}\left(\overline{\upsilon}''\right) = ([1.575, 2], [2.425, 4])^{\mathrm{T}},$$

which is accordant with Eqs. (3.35)–(3.38) and Eq. (3.41).

In the same way, for a given parameter δ ($\delta \neq 1$ or 0), if an interval-valued cooperative game $\overline{\upsilon} \in \overline{G}^n$ satisfies Eq. (3.31), then we can prove that its interval-valued discounted Shapley value $\overline{\boldsymbol{\Phi}}^{\mathrm{DSH}\delta}(\overline{\upsilon})$ satisfies the efficiency, the symmetry, the additivity, and the anonymity. However, $\overline{\boldsymbol{\Phi}}^{\mathrm{DSH}\delta}(\overline{\upsilon})$ does not always satisfy the invariance, the individual rationality, the dummy player property, and the null player property.

3.4 Interval-Valued Solidarity Values and Generalized Solidarity Values of Interval-Valued Cooperative Games

It is easy to see from the previous Sect. 3.2 and 3.3 that the Shapley value and the equal division value are two extreme cases of solution concepts of cooperative games [27, 28]. In fact, the Shapley value not only assigns zero payoffs to unproductive players (i.e., null players) but also may make these null players not to affect the other players' payoffs even if they leave the cooperative game [29]. Furthermore, a player's payoff only depends on his/her own marginal contributions. That is to say, the Shapley value does not allow for solidarity among the players [27, 30]. However, almost all modern societies reveal some degree of solidarity. In contrast to the Shapley value, the equal division value distributes the grand coalition's worth equally among the players. Apparently, a player' payoff allocated by the equal division value is almost insensitive to his/her own marginal contributions. Thus, the equal division value may be regarded as an extreme kind of solidarity. As a result, in 1994, Nowak and Radzik [28] proposed the solidarity value of a cooperative game. Later on, Casajus and Huettner [27] proposed the generalized solidarity value of a cooperative game. The solidarity value and the generalized solidarity value form an important one-parameter family of solidarity values for cooperative games [31]. In the following, we briefly review the concepts of the solidarity value and the generalized solidarity value.

For an arbitrary cooperative game $\upsilon \in G^n$ stated as in the previous Sect. 1.2, to define its solidarity value, we first introduce the average marginal contributions of the members in a coalition. To be more specific, for any coalition $S \subseteq N$, we define the average marginal contributions of the members in the coalition S as follows:

$$m(v,S) = \frac{1}{s}\sum_{j\in S}(v(S) - v(S\backslash j)). \tag{3.42}$$

Hereby, we can define the solidarity value $\boldsymbol{\rho}^{\mathrm{SV}}(v) = \left(\rho_1^{\mathrm{SV}}(v), \rho_2^{\mathrm{SV}}(v), \ldots, \rho_n^{\mathrm{SV}}(v)\right)^{\mathrm{T}}$ of the cooperative game v, whose components are given as follows:

$$\rho_i^{\mathrm{SV}}(v) = \sum_{S\subseteq N: i\in S} \frac{(s-1)!(n-s)!}{n!} m(v,S) \quad (i = 1, 2, \ldots, n). \tag{3.43}$$

Apparently, the solidarity value of a player $i \in S$ can be regarded as the weighted average of the average marginal contributions $m(v,S)$ where the player belongs to the coalition $S \subseteq N$.

It is easy to prove that the solidarity value $\boldsymbol{\rho}^{\mathrm{SV}}(v)$ of any cooperative game $v \in G^n$ satisfies the efficiency, the symmetry, and the additivity [28].

Similarly, we define the value $\boldsymbol{\rho}^{\mathrm{GSV}\xi}(v) = \left(\rho_1^{\mathrm{GSV}\xi}(v), \rho_2^{\mathrm{GSV}\xi}(v), \ldots, \rho_n^{\mathrm{GSV}\xi}(v)\right)^{\mathrm{T}}$ of a cooperative game $v \in G^n$, whose components are given as follows:

$$\begin{aligned}\rho_i^{\mathrm{GSV}\xi}(v) = \xi_n\frac{v(N)}{n} + \sum_{S\subseteq N\backslash i}\Bigg\{&\frac{s!(n-s-1)!}{n!} \\ &\times\left[(1-\xi_{s+1})v(S\cup i) - (1-\xi_s)v(S)\right]\Bigg\} \quad (i = 1, 2, \ldots, n),\end{aligned} \tag{3.44}$$

where

$$\xi_s = \frac{s\xi}{(s-1)\xi + 1} \quad (s = 0, 1, 2, \ldots, n) \tag{3.45}$$

and $\xi \in R\backslash\{-1/k | k \in N\}$, i.e., ξ is any real number but not equal to $-1/k$ $(k = 1, 2, \ldots, n)$. Particularly, if the parameter $\xi \in [0, 1]$, then $\boldsymbol{\rho}^{\mathrm{GSV}\xi}(v)$ is called the generalized solidarity value of the cooperative game $v \in G^n$.

Obviously, if $\xi = 0$, then the generalized solidarity value $\boldsymbol{\rho}^{\mathrm{GSV}\xi}(v)$ is reduced to the Shapley value $\boldsymbol{\Phi}^{\mathrm{SH}}(v)$, i.e., $\boldsymbol{\rho}^{\mathrm{GSV0}}(v) = \boldsymbol{\Phi}^{\mathrm{SH}}(v)$; if $\xi = 1$, then the generalized solidarity value $\boldsymbol{\rho}^{\mathrm{GSV}\xi}(v)$ is reduced to the equal division value $\boldsymbol{\rho}^{\mathrm{ED}}(v)$, i.e., $\boldsymbol{\rho}^{\mathrm{GSV1}}(v) = \boldsymbol{\rho}^{\mathrm{ED}}(v)$; if $\xi = 1/2$, then the generalized solidarity value $\boldsymbol{\rho}^{\mathrm{GSV}\xi}(v)$ is reduced to the solidarity value $\boldsymbol{\rho}^{\mathrm{SV}}(v)$, i.e., $\boldsymbol{\rho}^{\mathrm{GSV},0.5}(v) = \boldsymbol{\rho}^{\mathrm{SV}}(v)$. In the sequent, we often mean that $\boldsymbol{\rho}^{\mathrm{GSV}\xi}(v)$ is the generalized solidarity value of a cooperative game $v \in G^n$ unless otherwise stated.

We can easily check that the generalized solidarity value $\boldsymbol{\rho}^{\mathrm{GSV}\xi}(v)$ of any cooperative game $v \in G^n$ satisfies the efficiency, the symmetry, and the additivity [27].

In the following, to facilitate determining the condition of the monotonicity of the generalized solidarity value of a cooperative game, we need to give an alternative expression of Eq. (3.44). Firstly, we have

$$
\begin{aligned}
&-\sum_{S\subseteq N\backslash i}\frac{s!(n-s-1)!}{n!}[(1-\xi_{s+1})v(N)-(1-\xi_s)v(N)]\\
&=-\sum_{S\subseteq N\backslash i}\frac{s!(n-s-1)!}{n!}(\xi_s-\xi_{s+1})v(N)\\
&=v(N)\sum_{S\subseteq N\backslash i}\frac{s!(n-s-1)!}{n!}(\xi_{s+1}-\xi_s)\\
&=v(N)\sum_{k=0}^{n-1}\sum_{s=k}\frac{s!(n-s-1)!}{n!}(\xi_{s+1}-\xi_s)\\
&=v(N)\sum_{k=0}^{n-1}\frac{k!(n-k-1)!}{n!}(\xi_{k+1}-\xi_k)C_{n-1}^k\\
&=v(N)\sum_{k=0}^{n-1}\frac{k!(n-k-1)!}{n!}\times\frac{(n-1)!}{k!(n-k-1)!}(\xi_{k+1}-\xi_k)\\
&=v(N)\sum_{k=0}^{n-1}\frac{1}{n}(\xi_{k+1}-\xi_k)\\
&=\frac{v(N)}{n}(\xi_n-\xi_0),
\end{aligned}
$$

i.e.,

$$
-\sum_{S\subseteq N\backslash i}\frac{s!(n-s-1)!}{n!}[(1-\xi_{s+1})v(N)-(1-\xi_s)v(N)]=(\xi_n-\xi_0)\frac{v(N)}{n}.
$$

Combining with $\xi_0=0$ which is directly derived from Eq. (3.45), we easily obtain

$$
-\sum_{S\subseteq N\backslash i}\frac{s!(n-s-1)!}{n!}[(1-\xi_{s+1})v(N)-(1-\xi_s)v(N)]=\xi_n\frac{v(N)}{n}.
$$

Thus, Eq. (3.44) can be rewritten as follows:

$$
\begin{aligned}
\rho_i^{\mathrm{GSV}\xi}(v)=\sum_{S\subseteq N\backslash i}\Big\{&\frac{s!(n-s-1)!}{n!}[(1-\xi_{s+1})(v(S\cup i)-v(N))\\
&-(1-\xi_s)(v(S)-v(N))]\Big\}\quad(i=1,2,\ldots,n).
\end{aligned}
\tag{3.46}
$$

3.4.1 Interval-Valued Solidarity Values of Interval-Valued Cooperative Games and Simplified Methods

For an arbitrary interval-valued cooperative game $\overline{v} \in \overline{G}^n$ stated as in Sect. 1.3.2, we can similarly construct an associated cooperative game $v(\alpha) \in G^n$, where $N = \{1, 2, \ldots, n\}$ is the set of players and the characteristic function $v(\alpha)$ of coalitions of players is given by Eq. (3.4).

According to Eq. (3.43), we can easily obtain the solidarity value $\rho^{\mathrm{SV}}(v(\alpha)) = (\rho_1^{\mathrm{SV}}(v(\alpha)), \rho_2^{\mathrm{SV}}(v(\alpha)), \ldots, \rho_n^{\mathrm{SV}}(v(\alpha)))^{\mathrm{T}}$ of the cooperative game $v(\alpha) \in G^n$, whose components are given as follows:

$$\rho_i^{\mathrm{SV}}(v(\alpha)) = \sum_{S \subseteq N: i \in S} \frac{(s-1)!(n-s)!}{n!s} \sum_{j \in S} (v(\alpha)(S) - v(\alpha)(S \backslash j)) \quad (i = 1, 2, \ldots, n).$$

Combining with Eq. (3.4), we have

$$\rho_i^{\mathrm{SV}}(v(\alpha)) = \sum_{S \subseteq N: i \in S} \left\{ \frac{(s-1)!(n-s)!}{n!s} \times \sum_{j \in S} \{[(1-\alpha)v_L(S) + \alpha v_R(S)] - [(1-\alpha)v_L(S \backslash j) + \alpha v_R(S \backslash j)]\} \right\},$$

where $i = 1, 2, \ldots, n$. Thus, $\rho_i^{\mathrm{SV}}(v(\alpha))$ $((i = 1, 2, \ldots, n))$ can be further rewritten as follows:

$$\rho_i^{\mathrm{SV}}(v(\alpha)) = \sum_{S \subseteq N: i \in S} \left\{ \frac{(s-1)!(n-s)!}{n!s} \times \sum_{j \in S} [(1-\alpha)(v_L(S) - v_L(S \backslash j)) + \alpha(v_R(S) - v_R(S \backslash j))] \right\} \quad (i = 1, 2, \ldots, n). \tag{3.47}$$

Apparently, the solidarity value $\boldsymbol{\rho}^{\mathrm{SV}}(v(\alpha))$ of the cooperative game $v(\alpha) \in G^n$ is a continuous function of the parameter $\alpha \in [0, 1]$.

Theorem 3.28 *For any interval-valued cooperative game* $\overline{v} \in \overline{G}^n$, *if the following system of inequalities*

$$\sum_{S \subseteq N: i \in S} (v_R(S) - v_L(S)) \geq \sum_{S \subseteq N: i \in S} (v_R(S \backslash i) - v_L(S \backslash i)) \quad (i = 1, 2, \ldots, n) \tag{3.48}$$

is satisfied, then the solidarity value $\boldsymbol{\rho}^{\mathrm{SV}}(v(\alpha))$ *of the cooperative game* $v(\alpha) \in G^n$ *is a monotonic and non-decreasing function of the parameter* $\alpha \in [0, 1]$.

Proof For any $\alpha \in [0,1]$ and $\alpha' \in [0,1]$, if $\alpha \geq \alpha'$, according to Eq. (3.47), and combining with the assumption, i.e., Eq. (3.48), we have

$$\rho_i^{\mathrm{SV}}(v(\alpha)) - \rho_i^{\mathrm{SV}}(v(\alpha')) = \sum_{S \subseteq N: i \in S} \left\{ \frac{(s-1)!(n-s)!}{n!s} \times \sum_{j \in S} (\alpha - \alpha')[(v_R(S) - v_R(S \backslash j)) - (v_L(S) - v_L(S \backslash j))] \right\}$$
$$\geq 0,$$

where $i = 1, 2, \ldots, n$. Hence, we have

$$\rho_i^{\mathrm{SV}}(v(\alpha)) \geq \rho_i^{\mathrm{SV}}\left(v\left(\alpha'\right)\right) \quad (i = 1, 2, \ldots, n).$$

Thus, the solidarity value $\boldsymbol{\rho}^{\mathrm{SV}}(v(\alpha))$ is a monotonic and non-decreasing function of the parameter $\alpha \in [0,1]$. We have completed the proof of Theorem 3.28.

Thus, for any interval-valued cooperative game $\overline{v} \in \overline{G}^n$, if it satisfies Eq. (3.48), then according to Theorem 3.28, the lower and upper bounds of the interval-valued solidarity value $\overline{\boldsymbol{\rho}}^{\mathrm{SV}}(\overline{v}) = \left(\overline{\rho}_1^{\mathrm{SV}}(\overline{v}), \overline{\rho}_2^{\mathrm{SV}}(\overline{v}), \ldots, \overline{\rho}_n^{\mathrm{SV}}(\overline{v})\right)^{\mathrm{T}}$ can be attained at the lower and upper bounds of the interval [0, 1], respectively. Therefore, according to Eq. (3.47), we can directly and explicitly define the interval-valued solidarity value $\overline{\boldsymbol{\rho}}^{\mathrm{SV}}(\overline{v})$, whose components are given as follows:

$$\overline{\rho}_i^{\mathrm{SV}}(\overline{v}) = \left[\sum_{S \subseteq N: i \in S} \frac{(s-1)!(n-s)!}{n!s} \sum_{j \in S} (v_L(S) - v_L(S \backslash j)), \sum_{S \subseteq N: i \in S} \frac{(s-1)!(n-s)!}{n!s} \sum_{j \in S} (v_R(S) - v_R(S \backslash j)) \right] \quad (i = 1, 2, \ldots, n). \tag{3.49}$$

Example 3.10 Let us compute the interval-valued solidarity value $\overline{\boldsymbol{\rho}}^{\mathrm{SV}}\left(\overline{v}''\right)$ of the interval-valued cooperative game $\overline{v}'' \in \overline{G}^3$ given in Example 2.2.

From Example 2.2, we have $\overline{v}''(1,2) = \overline{v}''(1,3) = \overline{v}''\left(N''\right) = [291, 306]$ and $\overline{v}''(S) = 0$ for any other coalitions $S \subset N''$, where the grand coalition $N'' = \{1, 2, 3\}$. Thus, we easily have

$$\sum_{S \subseteq N'': 1 \in S} \left(v_R''(S) - v_L''(S)\right) = 45 > \sum_{S \subseteq N'': 1 \in S} \left(v_R''(S \backslash 1) - v_L''(S \backslash 1)\right) = 0,$$

$$\sum_{S\subseteq N'':2\in S}\left(v''_R(S)-v''_L(S)\right)=30>\sum_{S\subseteq N'':2\in S}\left(v''_R(S\backslash 2)-v''_L(S\backslash 2)\right)=15,$$

and

$$\sum_{S\subseteq N'':3\in S}\left(v''_R(S)-v''_L(S)\right)=30>\sum_{S\subseteq N'':3\in S}\left(v''_R(S\backslash 3)-v''_L(S\backslash 3)\right)=15,$$

i.e., the interval-valued cooperative game $\overline{v}''\in\overline{G}^3$ satisfies Eq. (3.48). Thus, according to Eq. (3.49), we can have

$$\begin{aligned}\rho^{SV}_{L1}(\overline{v}'')=&\sum_{S\subseteq N'':1\in S}\frac{(s-1)!(3-s)!}{3!s}\sum_{j\in S}\left(v''_L(S)-v''_L(S\backslash j)\right)\\
=&\frac{0!2!}{3!\times 1}\left(v''_L(1)-v''_L(\varnothing)\right)+\frac{1!1!}{3!\times 2}\left[\left(v''_L(1,2)-v''_L(2)\right)+\left(v''_L(1,2)-v''_L(1)\right)\right]\\
&+\frac{1!1!}{3!\times 2}\left[\left(v''_L(1,3)-v''_L(3)\right)+\left(v''_L(1,3)-v''_L(1)\right)\right]\\
&+\frac{2!0!}{3!\times 3}\left[\left(v''_L\left(N''\right)-v''_L(2,3)\right)+\left(v''_L\left(N''\right)-v''_L(1,3)\right)+\left(v''_L\left(N''\right)-v''_L(1,2)\right)\right]\\
=&\frac{1}{3}(0-0)+\frac{1}{12}[(291-0)+(291-0)]+\frac{1}{12}[(291-0)+(291-0)]\\
&+\frac{1}{9}[(291-0)+(291-291)+(291-291)]\\
=&\frac{1164}{9}\\
\approx&\,129.333\end{aligned}$$

and

$$\begin{aligned}\rho^{SV}_{R1}(\overline{v}'')=&\sum_{S\subseteq N'':1\in S}\frac{(s-1)!(3-s)!}{3!s}\sum_{j\in S}\left(v''_R(S)-v''_R(S\backslash j)\right)\\
=&\frac{0!2!}{3!\times 1}\left(v''_R(1)-v''_R(\varnothing)\right)+\frac{1!1!}{3!\times 2}\left[\left(v''_R(1,2)-v''_R(2)\right)+\left(v''_R(1,2)-v''_R(1)\right)\right]\\
&+\frac{1!1!}{3!\times 2}\left[\left(v''_R(1,3)-v''_R(3)\right)+\left(v''_R(1,3)-v''_R(1)\right)\right]\\
&+\frac{2!0!}{3!\times 3}\left[\left(v''_R\left(N''\right)-v''_R(2,3)\right)+\left(v''_R\left(N''\right)-v''_R(1,3)\right)+\left(v''_R\left(N''\right)-v''_R(1,2)\right)\right]\\
=&\frac{1}{3}(0-0)+\frac{1}{12}[(306-0)+(306-0)]+\frac{1}{12}[(306-0)+(306-0)]\\
&+\frac{1}{9}[(306-0)+(306-306)+(306-306)]\\
=&\,136.\end{aligned}$$

Analogously, we have

$$\begin{aligned}
\rho_{L2}^{SV}(\overline{v}'') &= \sum_{S\subseteq N'':2\in S} \frac{(s-1)!(3-s)!}{3!s} \sum_{j\in S} \left(v_L''(S) - v_L''(S\backslash j)\right) \\
&= \frac{0!2!}{3!\times 1}\left(v_L''(2) - v_L''(\varnothing)\right) + \frac{1!1!}{3!\times 2}\left[\left(v_L''(1,2) - v_L''(2)\right) + \left(v_L''(1,2) - v_L''(1)\right)\right] \\
&+ \frac{1!1!}{3!\times 2}\left[\left(v_L''(2,3) - v_L''(3)\right) + \left(v_L''(2,3) - v_L''(2)\right)\right] \\
&+ \frac{2!0!}{3!\times 3}\left[\left(v_L''\left(N''\right) - v_L''(2,3)\right) + \left(v_L''\left(N''\right) - v_L''(1,3)\right) + \left(v_L''\left(N''\right) - v_L''(1,2)\right)\right] \\
&= \frac{1}{3}(0-0) + \frac{1}{12}[(291-0) + (291-0)] + \frac{1}{12}[(0-0) + (0-0)] \\
&+ \frac{1}{9}[(291-0) + (291-291) + (291-291)] \\
&= \frac{1455}{18} \\
&\approx 80.833,
\end{aligned}$$

$$\begin{aligned}
\rho_{R2}^{SV}(\overline{v}'') &= \sum_{S\subseteq N'':2\in S} \frac{(s-1)!(3-s)!}{3!s} \sum_{j\in S} \left(v_R''(S) - v_R''(S\backslash j)\right) \\
&= \frac{0!2!}{3!\times 1}\left(v_R''(2) - v_R''(\varnothing)\right) + \frac{1!1!}{3!\times 2}\left[\left(v_R''(1,2) - v_R''(2)\right) + \left(v_R''(1,2) - v_R''(1)\right)\right] \\
&+ \frac{1!1!}{3!\times 2}\left[\left(v_R''(2,3) - v_R''(3)\right) + \left(v_R''(2,3) - v_R''(2)\right)\right] \\
&+ \frac{2!0!}{3!\times 3}\left[\left(v_R''\left(N''\right) - v_R''(2,3)\right) + \left(v_R''\left(N''\right) - v_R''(1,3)\right) + \left(v_R''\left(N''\right) - v_R''(1,2)\right)\right] \\
&= \frac{1}{3}(0-0) + \frac{1}{12}[(306-0) + (306-0)] + \frac{1}{12}[(0-0) + (0-0)] \\
&+ \frac{1}{9}[(306-0) + (306-306) + (306-306)] \\
&= 85,
\end{aligned}$$

$$\begin{aligned}
\rho_{L3}^{\mathrm{SV}}\left(\bar{v}''\right) &= \sum_{S \subseteq N'': 3 \in S} \frac{(s-1)!(3-s)!}{3!s} \sum_{j \in S} \left(v_L''(S) - v_L''(S \backslash j)\right) \\
&= \frac{0!2!}{3! \times 1}\left(v_L''(3) - v_L''(\varnothing)\right) + \frac{1!1!}{3! \times 2}\left[\left(v_L''(1,3) - v_L''(3)\right) + \left(v_L''(1,3) - v_L''(1)\right)\right] \\
&\quad + \frac{1!1!}{3! \times 2}\left[\left(v_L''(2,3) - v_L''(3)\right) + \left(v_L''(2,3) - v_L''(2)\right)\right] \\
&\quad + \frac{2!0!}{3! \times 3}\left[\left(v_L''\left(N''\right) - v_L''(2,3)\right) + \left(v_L''\left(N''\right) - v_L''(1,3)\right) + \left(v_L''\left(N''\right) - v_L''(1,2)\right)\right] \\
&= \frac{1}{3}(0-0) + \frac{1}{12}[(291-0) + (291-0)] + \frac{1}{12}[(0-0) + (0-0)] \\
&\quad + \frac{1}{9}[(291-0) + (291-291) + (291-291)] \\
&= \frac{1455}{18} \\
&\approx 80.833,
\end{aligned}$$

and

$$\begin{aligned}
\rho_{R3}^{\mathrm{SV}}\left(\bar{v}''\right) &= \sum_{S \subseteq N'': 3 \in S} \frac{(s-1)!(3-s)!}{3!s} \sum_{j \in S} \left(v_R''(S) - v_R''(S \backslash j)\right) \\
&= \frac{0!2!}{3! \times 1}\left(v_R''(3) - v_R''(\varnothing)\right) + \frac{1!1!}{3! \times 2}\left[\left(v_R''(1,3) - v_R''(3)\right) + \left(v_R''(1,3) - v_R''(1)\right)\right] \\
&\quad + \frac{1!1!}{3! \times 2}\left[\left(v_R''(2,3) - v_R''(3)\right) + \left(v_R''(2,3) - v_R''(2)\right)\right] \\
&\quad + \frac{2!0!}{3! \times 3}\left[\left(v_R''\left(N''\right) - v_R''(2,3)\right) + \left(v_R''\left(N''\right) - v_R''(1,3)\right) + \left(v_R''\left(N''\right) - v_R''(1,2)\right)\right] \\
&= \frac{1}{3}(0-0) + \frac{1}{12}[(306-0) + (306-0)] + \frac{1}{12}[(0-0) + (0-0)] \\
&\quad + \frac{1}{9}[(306-0) + (306-306) + (306-306)] \\
&= 85.
\end{aligned}$$

Therefore, the interval-valued solidarity value $\bar{\rho}^{\mathrm{SV}}\left(\bar{v}''\right)$ of the interval-valued cooperative game $\bar{v}'' \in \overline{G}^3$ can be obtained as follows:

$$\bar{\rho}^{\mathrm{SV}}\left(\bar{v}''\right) = \left(\left[\frac{1164}{9}, 136\right], \left[\frac{1455}{18}, 85\right], \left[\frac{1455}{18}, 85\right]\right)^{\mathrm{T}},$$

or approximately,

$$\bar{\rho}^{\mathrm{SV}}\left(\bar{v}''\right) = ([129.334, 136], [80.833, 85], [80.833, 85])^{\mathrm{T}},$$

which means that the player 1 (i.e., investor) gets the profit $[1164/9, 136]$ (or $[129.334, 136]$) and the players (i.e., technologists) 2 and 3 get the identical profit $[1455/18, 85]$ (or $[80.833, 85]$) from the cooperative production.

Obviously, the above interval-valued solidarity value $\overline{\rho}^{\mathrm{SV}}(\overline{\upsilon}'') = ([1164/9, 136], [1455/18, 85], [1455/18, 85])^{\mathrm{T}}$ is remarkably different from the interval-valued core $\overline{C}(\overline{\upsilon}'') = \left\{([291, 306], 0, 0)^{\mathrm{T}}\right\}$ given in Example 2.2.

In the same way, it is obvious that the interval-valued cooperative game $\overline{\upsilon}'' \in \overline{G}^3$ satisfies Eq. (3.17). Thus, according to Eqs. (3.18) and (3.19), we can easily have

$$\begin{aligned}
\phi_{L1}^{\mathrm{SH}}(\overline{\upsilon}'') &= \sum_{S \subseteq N'' \backslash 1} \frac{s!(2-s)!}{3!} \left(\upsilon_L''(S \cup 1) - \upsilon_L''(S)\right) \\
&= \frac{0!2!}{3!} \left(\upsilon_L''(1) - \upsilon_L''(\varnothing)\right) + \frac{1!1!}{3!} \left(\upsilon_L''(1,2) - \upsilon_L''(2)\right) + \frac{1!1!}{3!} \left(\upsilon_L''(1,3) - \upsilon_L''(3)\right) \\
&\quad + \frac{2!0!}{3!} \left(\upsilon_L''\left(N''\right) - \upsilon_L''(2,3)\right) \\
&= \frac{1}{3}(0-0) + \frac{1}{6}(291-0) + \frac{1}{6}(291-0) + \frac{1}{3}(291-0) \\
&= 194,
\end{aligned}$$

$$\begin{aligned}
\phi_{R1}^{\mathrm{SH}}(\overline{\upsilon}'') &= \sum_{S \subseteq N'' \backslash 1} \frac{s!(2-s)!}{3!} \left(\upsilon_R''(S \cup 1) - \upsilon_R''(S)\right) \\
&= \frac{0!2!}{3!} \left(\upsilon_R''(1) - \upsilon_R''(\varnothing)\right) + \frac{1!1!}{3!} \left(\upsilon_R''(1,2) - \upsilon_R''(2)\right) + \frac{1!1!}{3!} \left(\upsilon_R''(1,3) - \upsilon_R''(3)\right) \\
&\quad + \frac{2!0!}{3!} \left(\upsilon_R''\left(N''\right) - \upsilon_R''(2,3)\right) \\
&= \frac{1}{3}(0-0) + \frac{1}{6}(306-0) + \frac{1}{6}(306-0) + \frac{1}{3}(306-0) \\
&= 204,
\end{aligned}$$

$$\begin{aligned}
\phi_{L2}^{\mathrm{SH}}(\overline{\upsilon}'') &= \sum_{S \subseteq N'' \backslash 2} \frac{s!(2-s)!}{3!} \left(\upsilon_L''(S \cup 2) - \upsilon_L''(S)\right) \\
&= \frac{0!2!}{3!} \left(\upsilon_L''(2) - \upsilon_L''(\varnothing)\right) + \frac{1!1!}{3!} \left(\upsilon_L''(1,2) - \upsilon_L''(1)\right) + \frac{1!1!}{3!} \left(\upsilon_L''(2,3) - \upsilon_L''(3)\right) \\
&\quad + \frac{2!0!}{3!} \left(\upsilon_L''\left(N''\right) - \upsilon_L''(1,3)\right) \\
&= \frac{1}{3}(0-0) + \frac{1}{6}(291-0) + \frac{1}{6}(0-0) + \frac{1}{3}(291-291) \\
&= 48.5,
\end{aligned}$$

$$\phi_{R2}^{\mathrm{SH}}(\overline{v}'') = \sum_{S \subseteq N'' \setminus 2} \frac{s!(2-s)!}{3!}\left(v_R''(S \cup 2) - v_R''(S)\right)$$
$$= \frac{0!2!}{3!}\left(v_R''(2) - v_R''(\varnothing)\right) + \frac{1!1!}{3!}\left(v_R''(1,2) - v_R''(1)\right) + \frac{1!1!}{3!}\left(v_R''(2,3) - v_R''(3)\right)$$
$$+ \frac{2!0!}{3!}\left(v_R''\left(N''\right) - v_R''(1,3)\right)$$
$$= \frac{1}{3}(0-0) + \frac{1}{6}(306-0) + \frac{1}{6}(0-0) + \frac{1}{3}(306-306)$$
$$= 51,$$

$$\phi_{L3}^{\mathrm{SH}}(\overline{v}'') = \sum_{S \subseteq N'' \setminus 3} \frac{s!(2-s)!}{3!}\left(v_L''(S \cup 3) - v_L''(S)\right)$$
$$= \frac{0!2!}{3!}\left(v_L''(3) - v_L''(\varnothing)\right) + \frac{1!1!}{3!}\left(v_L''(1,3) - v_L''(1)\right) + \frac{1!1!}{3!}\left(v_L''(2,3) - v_L''(2)\right)$$
$$+ \frac{2!0!}{3!}\left(v_L''\left(N''\right) - v_L''(1,2)\right)$$
$$= \frac{1}{3}(0-0) + \frac{1}{6}(291-0) + \frac{1}{6}(0-0) + \frac{1}{3}(291-291)$$
$$= 48.5,$$

and

$$\phi_{R3}^{\mathrm{SH}}(\overline{v}'') = \sum_{S \subseteq N'' \setminus 3} \frac{s!(2-s)!}{3!}\left(v_R''(S \cup 3) - v_R''(S)\right)$$
$$= \frac{0!2!}{3!}\left(v_R''(3) - v_R''(\varnothing)\right) + \frac{1!1!}{3!}\left(v_R''(1,3) - v_R''(1)\right) + \frac{1!1!}{3!}\left(v_R''(2,3) - v_R''(2)\right)$$
$$+ \frac{2!0!}{3!}\left(v_R''\left(N''\right) - v_R''(1,2)\right)$$
$$= \frac{1}{3}(0-0) + \frac{1}{6}(306-0) + \frac{1}{6}(0-0) + \frac{1}{3}(306-306)$$
$$= 51.$$

Therefore, the interval-valued Shapley value $\overline{\boldsymbol{\Phi}}^{\mathrm{SH}}(\overline{v}'')$ of the above interval-valued cooperative game $\overline{v}'' \in \overline{G}^3$ is obtained as follows:

$$\overline{\boldsymbol{\Phi}}^{\mathrm{SH}}\left(\overline{v}''\right) = ([194, 204], [48.5, 51], [48.5, 51])^{\mathrm{T}},$$

which is different from the interval-valued solidarity value $\overline{\boldsymbol{\rho}}^{\mathrm{SV}}(\overline{v}'')$ also. The main difference is that the player 1 (i.e., investor) gets considerably more profit from the interval-valued Shapley value $\overline{\boldsymbol{\Phi}}^{\mathrm{SH}}(\overline{v}'')$ and the interval-valued core $\overline{C}(\overline{v}'')$ than that from the interval-valued solidarity value $\overline{\boldsymbol{\rho}}^{\mathrm{SV}}(\overline{v}'')$.

Equation (3.48) can be rewritten as follows:

$$\sum_{S\subseteq N:i\in S} l(S) \geq \sum_{S\subseteq N:i\in S} l(S\backslash i) \quad (i = 1, 2, \ldots, n),$$

which mean that the sum of the value (i.e., interval) lengths of the coalitions including the player $i \in S$ is monotonic. This kind of the monotonicity is related to coalitions. Thus, in a similar way to Sect. 3.3.1, if an interval-valued cooperative game $\overline{v} \in \overline{G}^n$ satisfies Eq. (3.48), then it is called size monotonic-like.

Obviously, the condition given by Eq. (3.48) is weaker than that given by Eq. (3.17). In other word, if Eq. (3.17) is satisfied, i.e.,

$$v_R(S) - v_L(S) \geq v_R(S\backslash i) - v_L(S\backslash i) \quad (i \in S \subseteq N; i = 1, 2, \ldots, n),$$

then Eq. (3.48) is always valid. Conversely, i.e., Eq. (3.48) is satisfied, but Eq. (3.17) is not always true. This case can be illustrated with the following simple example.

Example 3.11 Let us discuss the interval-valued cooperative game $\overline{v}^0 \in \overline{G}^2$ with the interval-valued characteristic function as follows: $\overline{v}^0(1) = [2, 3.5]$, $\overline{v}^0(2) = [3, 5]$, $\overline{v}^0(1, 2) = [5, 6]$, and $\overline{v}^0(\emptyset) = 0$, where $N^0 = \{1, 2\}$.

It is obvious from the above interval-valued characteristic function $\overline{v}^0$ that

$$\sum_{S\subseteq N^0:1\in S} (v_R^0(S) - v_L^0(S)) = (3.5 - 2) + (6 - 5) = 2.5,$$

$$\sum_{S\subseteq N^0:1\in S} (v_R^0(S\backslash 1) - v_L^0(S\backslash 1)) = (0 - 0) + (5 - 3) = 2,$$

$$\sum_{S\subseteq N^0:2\in S} (v_R^0(S) - v_L^0(S)) = (5 - 3) + (6 - 5) = 3,$$

and

$$\sum_{S\subseteq N^0:2\in S} (v_R^0(S\backslash 2) - v_L^0(S\backslash 2)) = (0 - 0) + (3.5 - 2) = 1.5.$$

Hence, we have

$$\sum_{S\subseteq N^0:1\in S} (v_R^0(S) - v_L^0(S)) > \sum_{S\subseteq N^0:1\in S} (v_R^0(S\backslash 1) - v_L^0(S\backslash 1))$$

and

$$\sum_{S\subseteq N^0:2\in S} (v_R^0(S) - v_L^0(S)) > \sum_{S\subseteq N^0:2\in S} (v_R^0(S\backslash 2) - v_L^0(S\backslash 2)).$$

That is to say, the interval-valued cooperative game $\overline{\upsilon}^0 \in \overline{G}^2$ satisfies Eq. (3.48). But, it is easy to see that

$$\upsilon_R^0(1,2) - \upsilon_L^0(1,2) = 6 - 5 = 1 < \upsilon_R^0(1) - \upsilon_L^0(1) = 3.5 - 2 = 1.5$$

and

$$\upsilon_R^0(1,2) - \upsilon_L^0(1,2) = 6 - 5 = 1 < \upsilon_R^0(2) - \upsilon_L^0(2) = 5 - 3 = 2.$$

Namely, the interval-valued cooperative game $\overline{\upsilon}^0 \in \overline{G}^2$ does not satisfy Eq. (3.17).

In the sequent, we study some useful and important properties of interval-valued solidarity values of interval-valued cooperative games.

Theorem 3.29 *(Existence and Uniqueness) For an arbitrary interval-valued cooperative game $\overline{\upsilon} \in \overline{G}^n$, if it satisfies Eq. (3.48), then there always exists a unique interval-valued solidarity value $\overline{\boldsymbol{\rho}}^{\mathrm{SV}}(\overline{\upsilon})$, which is determined by Eq. (3.49).*

Proof According to Eq. (3.49), and combining with Definition 1.1, we can easily prove Theorem 3.29.

Theorem 3.30 *(Efficiency) For any interval-valued cooperative game $\overline{\upsilon} \in \overline{G}^n$, if it satisfies Eq. (3.48), then its interval-valued solidarity value $\overline{\boldsymbol{\rho}}^{\mathrm{SV}}(\overline{\upsilon})$ satisfies the efficiency, i.e., $\sum_{i=1}^{n} \overline{\rho}_i^{\mathrm{SV}}(\overline{\upsilon}) = \overline{\upsilon}(N)$.*

Proof According to Eq. (3.49), we can easily prove Theorem 3.30 in the same way to the proof given by Nowak and Radzik [28] (omitted).

Theorem 3.31 *(Additivity) For any interval-valued cooperative games $\overline{\upsilon} \in \overline{G}^n$ and $\overline{\nu} \in \overline{G}^n$, if they satisfy Eq. (3.48), then $\overline{\boldsymbol{\rho}}^{\mathrm{SV}}(\overline{\upsilon} + \overline{\nu}) = \overline{\boldsymbol{\rho}}^{\mathrm{SV}}(\overline{\upsilon}) + \overline{\boldsymbol{\rho}}^{\mathrm{SV}}(\overline{\nu})$.*

Proof According to Eq. (3.49) and Definition 1.1, we have

$$\overline{\rho}_i^{SV}(\overline{\upsilon}+\overline{\nu}) = \left[\sum_{S\subseteq N:i\in S} \frac{(s-1)!(n-s)!}{n!s} \sum_{j\in S} [(\upsilon_L(S)+\nu_L(S)) - (\upsilon_L(S\backslash j)+\nu_L(S\backslash j))],\right.$$

$$\left.\sum_{S\subseteq N:i\in S} \frac{(s-1)!(n-s)!}{n!s} \sum_{j\in S} [(\upsilon_R(S)+\nu_R(S)) - (\upsilon_R(S\backslash j)+\nu_L(S\backslash j))]\right]$$

$$= \left[\sum_{S\subseteq N:i\in S} \frac{(s-1)!(n-s)!}{n!s} \sum_{j\in S} (\upsilon_L(S)-\upsilon_L(S\backslash j)),\right.$$

$$\left.\sum_{S\subseteq N:i\in S} \frac{(s-1)!(n-s)!}{n!s} \sum_{j\in S} (\upsilon_R(S)-\upsilon_R(S\backslash j))\right]$$

$$+ \left[\sum_{S\subseteq N:i\in S} \frac{(s-1)!(n-s)!}{n!s} \sum_{j\in S} (\nu_L(S)-\nu_L(S\backslash j)),\right.$$

$$\left.\sum_{S\subseteq N:i\in S} \frac{(s-1)!(n-s)!}{n!s} \sum_{j\in S} (\nu_R(S)-\nu_L(S\backslash j))\right]$$

$$= \overline{\rho}_i^{SV}(\overline{\upsilon}) + \overline{\rho}_i^{SV}(\overline{\nu}),$$

i.e.,

$$\overline{\rho}_i^{SV}(\overline{\upsilon}+\overline{\nu}) = \overline{\rho}_i^{SV}(\overline{\upsilon}) + \overline{\rho}_i^{SV}(\overline{\nu}) \quad (i = 1, 2, \ldots, n).$$

Thus, we can obtain

$$\overline{\boldsymbol{\rho}}^{SV}(\overline{\upsilon}+\overline{\nu}) = \overline{\boldsymbol{\rho}}^{SV}(\overline{\upsilon}) + \overline{\boldsymbol{\rho}}^{SV}(\overline{\nu}).$$

Therefore, we have completed the proof of Theorem 3.31.

Theorem 3.32 *(Symmetry) For any interval-valued cooperative game $\overline{\upsilon} \in \overline{G}^n$, if it satisfies Eq. (3.48), and players $i \in N$ and $k \in N$ $(i \neq k)$ are symmetric in the interval-valued cooperative game $\overline{\upsilon}$, then $\overline{\rho}_i^{SV}(\overline{\upsilon}) = \overline{\rho}_k^{SV}(\overline{\upsilon})$.*

Proof Due to the assumption that the players $i \in N$ and $k \in N (i \neq k)$ are symmetric in the interval-valued cooperative game $\overline{\upsilon} \in \overline{G}^n$, then according to Definition 1.3, we have

$$\overline{\upsilon}(S \cup i) = \overline{\upsilon}(S \cup k),$$

i.e.,

$$\upsilon_L(S \cup i) = \upsilon_L(S \cup k)$$

and

$$\upsilon_R(S \cup i) = \upsilon_R(S \cup k).$$

According to Eq. (3.49), we can easily check that $\overline{\rho}_i^{SV}(\overline{\upsilon}) = \overline{\rho}_k^{SV}(\overline{\upsilon})$. Thus, we have completed the proof of Theorem 3.32.

Theorem 3.33 *(Anonymity) For any interval-valued cooperative game* $\bar{v} \in \overline{G}^n$ *and any permutation* σ *on the set N, if* $\bar{v}$ *satisfies Eq. (3.48), then* $\bar{\rho}^{\mathrm{SV}}_{\sigma(i)}(\bar{v}^\sigma) = \bar{\rho}^{\mathrm{SV}}_i(\bar{v})$ $(i = 1, 2, \ldots, n)$. *Namely,* $\bar{\rho}^{\mathrm{SV}}(\bar{v}^\sigma) = \sigma^{\#}(\bar{\rho}^{\mathrm{SV}}(\bar{v}))$.

Proof According to Eq. (3.49), Theorem 3.33 can be easily proved in a similar way to that of Theorem 3.32 (omitted).

Note that interval-valued solidarity values of interval-valued cooperative games do not always satisfy the invariance and the individual rationality.

3.4.2 Interval-Valued Generalized Solidarity Values of Interval-Valued Cooperative Games and Properties

As stated earlier, for the associated cooperative game $v(\alpha) \in G^n$ of an arbitrary interval-valued cooperative game $\bar{v} \in \overline{G}^n$ stated as in Sect. 1.3.2, according to Eq. (3.44) or Eq. (3.46), we can easily obtain its generalized solidarity value as follows:

$$\rho^{\mathrm{GSV}\xi}(v(\alpha)) = \left(\rho_1^{\mathrm{GSV}\xi}(v(\alpha)), \rho_2^{\mathrm{GSV}\xi}(v(\alpha)), \ldots, \rho_n^{\mathrm{GSV}\xi}(v(\alpha))\right)^{\mathrm{T}},$$

whose components are given as follows:

$$\rho_i^{\mathrm{GSV}\xi}(v(\alpha)) = \xi_n \frac{v(\alpha)(N)}{n} + \sum_{S \subseteq N \backslash i} \left\{ \frac{s!(n-s-1)!}{n!} \times [(1-\xi_{s+1})v(\alpha)(S \cup i) - (1-\xi_s)v(\alpha)(S)] \right\} \quad (i = 1, 2, \ldots, n) \tag{3.50}$$

or

$$\rho_i^{\mathrm{GSV}\xi}(v(\alpha)) = \sum_{S \subseteq N \backslash i} \left\{ \frac{s!(n-s-1)!}{n!} [(1-\xi_{s+1})(v(\alpha)(S \cup i) - v(\alpha)(N)) - (1-\xi_s)(v(\alpha)(S) - v(\alpha)(N))] \right\} \quad (i = 1, 2, \ldots, n), \tag{3.51}$$

where ξ_s $(s = 0, 1, 2, \ldots, n)$ are given by Eq. (3.45). It is worthwhile to note that ξ_s is a monotonic and non-decreasing function of the variable s. Namely,

$$\xi_{s+1} \geq \xi_s$$

for $s = 0, 1, 2, \ldots, n$.

Clearly, the generalized solidarity value $\rho^{\mathrm{GSV}\xi}(v(\alpha))$ of the cooperative game $v(\alpha) \in G^n$ is a continuous function of the parameter $\alpha \in [0, 1]$.

Theorem 3.34 *For any interval-valued cooperative game $\overline{v} \in \overline{G}^n$ and a given parameter $\xi \in [0, 1]$, if the following system of inequalities*

$$\begin{aligned}(1-\xi_{s+1})(v_R(S\cup i)-v_L(S\cup i))+(\xi_{s+1}-\xi_s)(v_R(N)-v_L(N))\\ \geq(1-\xi_s)(v_R(S)-v_L(S))\quad(S\subseteq N\backslash i; i=1,2,\ldots,n)\end{aligned} \tag{3.52}$$

is satisfied, then the generalized solidarity value $\boldsymbol{\rho}^{\mathrm{GSV}\xi}(v(\alpha))$ of the cooperative game $v(\alpha) \in G^n$ is a monotonic and non-decreasing function of the parameter $\alpha \in [0, 1]$.

Proof For any $\alpha \in [0, 1]$ and $\alpha' \in [0, 1]$, according to Eq. (3.51), and combining with Eq. (3.4), we have

$$\begin{aligned}\rho_i^{\mathrm{GSV}\xi}(v(\alpha))-\rho_i^{\mathrm{GSV}\xi}(v(\alpha'))=(\alpha-\alpha')\sum_{S\subseteq N\backslash i}\Bigg\{\frac{s!(n-s-1)!}{n!}\\ \times\{(1-\xi_{s+1})[(v_R(S\cup i)-v_L(S\cup i))-(v_R(N)-v_L(N))]\\ -(1-\xi_s)[(v_R(S)-v_L(S))-(v_R(N)-v_L(N))]\}\Bigg\},\end{aligned}$$

where $i = 1, 2, \ldots, n$. If $\alpha \geq \alpha'$, then combining with the assumption, i.e., Eq. (3.52), we have

$$\rho_i^{\mathrm{GSV}\xi}(v(\alpha))-\rho_i^{\mathrm{GSV}\xi}\left(v\left(\alpha'\right)\right)\geq 0\quad(i=1,2,\ldots,n),$$

which directly infer that

$$\rho_i^{\mathrm{SV}}(v(\alpha))\geq\rho_i^{\mathrm{SV}}\left(v\left(\alpha'\right)\right)\quad(i=1,2,\ldots,n).$$

Thus, the generalized solidarity value $\boldsymbol{\rho}^{\mathrm{GSV}\xi}(v(\alpha))$ is a monotonic and non-decreasing function of the parameter $\alpha \in [0, 1]$. We have completed the proof of Theorem 3.34.

Therefore, for any interval-valued cooperative game $\overline{v} \in \overline{G}^n$, if it satisfies Eq. (3.52), then according to Theorem 3.34, the lower and upper bounds of the interval-valued generalized solidarity value $\overline{\boldsymbol{\rho}}^{\mathrm{GSV}\xi}(\overline{v}) = \left(\overline{\rho}_1^{\mathrm{GSV}\xi}(\overline{v}), \overline{\rho}_2^{\mathrm{GSV}\xi}(\overline{v}), \ldots, \overline{\rho}_n^{\mathrm{GSV}\xi}(\overline{v})\right)^{\mathrm{T}}$ can be attained at the lower and upper bounds of the interval [0, 1], respectively. Thus, according to Eq. (3.50) or Eq. (3.51), we can directly and explicitly define the interval-valued generalized solidarity value $\overline{\boldsymbol{\rho}}^{\mathrm{GSV}\xi}(\overline{v})$, whose components are given as follows:

$$\bar{\rho}_i^{\mathrm{GSV}\xi}(\bar{v}) = \Bigg[\xi_n \frac{v_L(N)}{n} + \sum_{S\subseteq N\backslash i} \frac{s!(n-s-1)!}{n!}[(1-\xi_{s+1})v_L(S\cup i) - (1-\xi_s)v_L(S)],$$
$$\xi_n \frac{v_R(N)}{n} + \sum_{S\subseteq N\backslash i}\Bigg\{\frac{s!(n-s-1)!}{n!}[(1-\xi_{s+1})v_R(S\cup i)$$
$$-(1-\xi_s)v_R(S)]\Bigg\}\Bigg] \quad (i = 1,2,\ldots,n) \tag{3.53}$$

or

$$\bar{\rho}_i^{\mathrm{GSV}\xi}(\bar{v}) = \Bigg[\sum_{S\subseteq N\backslash i} \frac{s!(n-s-1)!}{n!}[(1-\xi_{s+1})(v_L(S\cup i) - v_L(N)) - (1-\xi_s)(v_L(S) - v_L(N))],$$
$$\sum_{S\subseteq N\backslash i}\Bigg\{\frac{s!(n-s-1)!}{n!}[(1-\xi_{s+1})(v_R(S\cup i) - v_R(N))$$
$$-(1-\xi_s)(v_R(S) - v_R(N))]\Bigg\}\Bigg] \quad (i = 1,2,\ldots,n). \tag{3.54}$$

Apparently, if $\xi = 0$, then the interval-valued generalized solidarity value $\bar{\boldsymbol{\rho}}^{\mathrm{GSV}\xi}(\bar{v})$ is reduced to the interval-valued Shapley value $\bar{\boldsymbol{\Phi}}^{\mathrm{SH}}(\bar{v})$, i.e., $\bar{\boldsymbol{\rho}}^{\mathrm{GSV0}}(\bar{v}) = \bar{\boldsymbol{\Phi}}^{\mathrm{SH}}(\bar{v})$; if $\xi = 1$, then interval-valued generalized solidarity value $\bar{\boldsymbol{\rho}}^{\mathrm{GSV}\xi}(\bar{v})$ is reduced to the interval-valued equal division value $\bar{\boldsymbol{\rho}}^{\mathrm{ED}}(\bar{v})$, i.e., $\bar{\boldsymbol{\rho}}^{\mathrm{GSV1}}(\bar{v}) = \bar{\boldsymbol{\rho}}^{\mathrm{ED}}(\bar{v})$; if $\xi = 1/2$, then interval-valued generalized solidarity value $\bar{\boldsymbol{\rho}}^{\mathrm{GSV}\xi}(\bar{v})$ is reduced to the interval-valued solidarity value $\bar{\boldsymbol{\rho}}^{\mathrm{SV}}(\bar{v})$, i.e., $\bar{\boldsymbol{\rho}}^{\mathrm{GSV},0.5}(\bar{v}) = \bar{\boldsymbol{\rho}}^{\mathrm{SV}}(\bar{v})$.

For an arbitrary interval-valued cooperative game $\bar{v} \in \bar{G}^n$ and a given parameter $\xi \in [0, 1]$, if $\bar{v}$ satisfies Eq. (3.52), then its interval-valued generalized solidarity value $\bar{\boldsymbol{\rho}}^{\mathrm{GSV}\xi}(\bar{v})$ possesses some useful and important properties, which are respectively summarized as in Theorems 3.35–3.39 as follows.

Theorem 3.35 *(Existence and Uniqueness) For an arbitrary interval-valued cooperative game $\bar{v} \in \bar{G}^n$ and a given parameter $\xi \in [0, 1]$, if $\bar{v}$ satisfies Eq. (3.52), then there always exists a unique interval-valued generalized solidarity value $\bar{\boldsymbol{\rho}}^{\mathrm{GSV}\xi}(\bar{v})$, which is determined by Eq. (3.53) or Eq. (3.54).*

Theorem 3.36 *(Efficiency) For any interval-valued cooperative game $\bar{v} \in \bar{G}^n$ and a given parameter $\xi \in [0, 1]$, if $\bar{v}$ satisfies Eq. (3.52), then its interval-valued generalized solidarity value $\bar{\boldsymbol{\rho}}^{\mathrm{GSV}\xi}(\bar{v})$ satisfies the efficiency, i.e., $\sum_{i=1}^n \bar{\rho}_i^{\mathrm{GSV}\xi}(\bar{v}) = \bar{v}(N)$.*

Theorem 3.37 *(Additivity) For any two interval-valued cooperative games* $\overline{v} \in \overline{G}^n$ *and* $\overline{\nu} \in \overline{G}^n$ *and a given parameter* $\xi \in [0,1]$, *if* $\overline{v}$ *and* $\overline{\nu}$ *satisfy Eq. (3.52), then* $\overline{\rho}^{\mathrm{GSV}\xi}(\overline{v}+\overline{\nu}) = \overline{\rho}^{\mathrm{GSV}\xi}(\overline{v}) + \overline{\rho}^{\mathrm{GSV}\xi}(\overline{\nu})$.

Theorem 3.38 *(Symmetry) For any interval-valued cooperative game* $\overline{v} \in \overline{G}^n$ *and a given parameter* $\xi \in [0,1]$, *if* $\overline{v}$ *satisfies Eq. (3.52), and players* $i \in N$ *and* $k \in N$ $(i \neq k)$ *are symmetric in the interval-valued cooperative game* $\overline{v}$, *then* $\overline{\rho}_i^{\mathrm{GSV}\xi}(\overline{v}) = \overline{\rho}_k^{\mathrm{GSV}\xi}(\overline{v})$.

Theorem 3.39 *(Anonymity) For any interval-valued cooperative game* $\overline{v} \in \overline{G}^n$, *any permutation* σ *on the set* N, *and a given parameter* $\xi \in [0,1]$, *if* $\overline{v}$ *satisfies Eq. (3.52), then* $\overline{\rho}_{\sigma(i)}^{\mathrm{GSV}\xi}(\overline{v}^{\sigma}) = \overline{\rho}_i^{\mathrm{GSV}\xi}(\overline{v})$ $(i = 1, 2, \ldots, n)$. *Namely,* $\overline{\rho}^{\mathrm{GSV}\xi}(\overline{v}^{\sigma}) = \sigma^{\#}(\overline{\rho}^{\mathrm{GSV}\xi}(\overline{v}))$.

According to Eq. (3.53) or Eq. (3.54), in a similar way to Theorems 3.29–3.33, we can easily complete the proof of Theorems 3.35–3.39, respectively (omitted).

Analogously, it is easy to see that interval-valued generalized solidarity values of interval-valued cooperative games do not always satisfy the invariance and the individual rationality.

3.5 Interval-Valued Banzhaf Values of Interval-Valued Cooperative Games

In 1965, Banzhaf [32] introduced an important index which is used as measures of power in cooperative games. This index usually is called the Banzhaf value or the Banzhaf index [33], which currently has various variants [34, 35]. In classical cooperative game theory, for an arbitrary cooperative game $v \in G^n$ stated as in the previous Sect. 1.2, we define its Banzhaf value as follows:

$$\rho^{\mathrm{B}}(v) = \left(\rho_1^{\mathrm{B}}(v), \rho_2^{\mathrm{B}}(v), \ldots, \rho_n^{\mathrm{B}}(v)\right)^{\mathrm{T}},$$

whose components are given as follows:

$$\rho_i^{\mathrm{B}}(v) = \frac{1}{2^{n-1}} \sum_{S \subseteq N \setminus i} (v(S \cup i) - v(S)) \quad (i = 1, 2, \ldots, n). \tag{3.55}$$

In a parallel way, based on the special interval subtraction [19], i.e.,

$$\overline{a} - \overline{b} = [a_L - b_L, a_R - b_R] \text{ if } a_R - a_L \geq b_R - b_L,$$

where $\overline{a} = [a_L, a_R] \in \overline{R}$ and $\overline{b} = [b_L, b_R] \in \overline{R}$, Pusillo [36] defined the Banzhaf-like value for a size monotonic interval-valued cooperative game. However, apparently, the defined Banzhaf like value is only applicable to a special class of interval-valued cooperative games. In the following, in the same way to the previous

Sects. 3.2–3.4, we will define the interval-valued Banzhaf value for an arbitrary interval-valued cooperative game $\bar{v} \in \overline{G}^n$ stated as in Sect. 1.3.2.

Analogously, for an arbitrary interval-valued cooperative game $\bar{v} \in \overline{G}^n$, we can construct an associated cooperative game $v(\alpha) \in G^n$, where the set of players is $N = \{1, 2, \ldots, n\}$ and the characteristic function $v(\alpha)$ of coalitions of players is given by Eq. (3.4).

According to Eq. (3.55), we can easily obtain the Banzhaf value $\boldsymbol{\rho}^{\mathrm{B}}(v(\alpha)) = \left(\rho_1^{\mathrm{B}}(v(\alpha)), \rho_2^{\mathrm{B}}(v(\alpha)), \ldots, \rho_n^{\mathrm{B}}(v(\alpha))\right)^{\mathrm{T}}$ of the cooperative game $v(\alpha) \in G^n$, where

$$\rho_i^{\mathrm{B}}(v(\alpha)) = \frac{1}{2^{n-1}} \sum_{S \subseteq N \backslash i} (v(\alpha)(S \cup i) - v(\alpha)(S)) \quad (i = 1, 2, \ldots, n),$$

which can further be rewritten as follows:

$$\begin{aligned} \rho_i^{\mathrm{B}}(v(\alpha)) = \frac{1}{2^{n-1}} \sum_{S \subseteq N \backslash i} \{[(1-\alpha)v_L(S \cup i) + \alpha v_R(S \cup i)] \\ - [(1-\alpha)v_L(S) + \alpha v_R(S)]\} \quad (i = 1, 2, \ldots, n). \end{aligned} \tag{3.56}$$

Obviously, the Banzhaf value $\boldsymbol{\rho}^{\mathrm{B}}(v(\alpha))$ is a continuous function of the parameter $\alpha \in [0, 1]$.

Theorem 3.40 *For any interval-valued cooperative game $\bar{v} \in \overline{G}^n$, if it satisfies Eq. (3.17), then the Banzhaf value $\boldsymbol{\rho}^{\mathrm{B}}(v(\alpha))$* of the cooperative game $v(\alpha) \in G^n$ is a monotonic and non-decreasing function of the parameter $\alpha \in [0, 1]$.

Proof For any $\alpha \in [0, 1]$ and $\alpha' \in [0, 1]$, according to Eq. (3.56), and using Eq. (3.4), we have

$$\begin{aligned} \rho_i^{\mathrm{B}}(v(\alpha)) - \rho_i^{\mathrm{B}}(v(\alpha')) = & \frac{1}{2^{n-1}} \sum_{S \subseteq N \backslash i} \{[(1-\alpha)v_L(S \cup i) + \alpha v_R(S \cup i)] \\ & - [(1-\alpha)v_L(S) + \alpha v_R(S)]\} - \frac{1}{2^{n-1}} \sum_{S \subseteq N \backslash i} \left\{\left[\left(1-\alpha'\right)v_L(S \cup i)\right.\right. \\ & \left.\left. + \alpha' v_R(S \cup i)\right] - \left[(1-\alpha')v_L(S) + \alpha' v_R(S)\right]\right\} \\ = & \frac{1}{2^{n-1}}\left(\alpha - \alpha'\right) \sum_{S \subseteq N \backslash i} \left[(v_R(S \cup i) - v_L(S \cup i)) - (v_R(S) - v_L(S))\right], \end{aligned}$$

where $i = 1, 2, \ldots, n$.

If $\alpha \geq \alpha'$, then combining with the assumption, i.e., Eq. (3.17), we have

$$\rho_i^{\mathrm{B}}(v(\alpha)) - \rho_i^{\mathrm{B}}\left(v\left(\alpha'\right)\right) \geq 0 \quad (i = 1, 2, \ldots, n),$$

i.e.,

$$\rho_i^{\mathrm{B}}(v(\alpha)) \geq \rho_i^{\mathrm{B}}\left(v\left(\alpha'\right)\right) \quad (i = 1, 2, \ldots, n),$$

which mean that the Banzhaf value $\boldsymbol{\rho}^{\mathrm{B}}(v(\alpha))$ is a monotonic and non-decreasing function of the parameter $\alpha \in [0, 1]$. Thus, we have completed the proof of Theorem 3.40.

Thus, for any interval-valued cooperative game $\overline{v} \in \overline{G}^n$, if it satisfies Eq. (3.17), then according to Theorem 3.40 and Eq. (3.56), we can directly and explicitly define its interval-valued Banzhaf value as follows:

$$\overline{\boldsymbol{\rho}}^{\mathrm{B}}(\overline{v}) = \left(\overline{\rho}_1^{\mathrm{B}}(\overline{v}), \overline{\rho}_2^{\mathrm{B}}(\overline{v}), \ldots, \overline{\rho}_n^{\mathrm{B}}(\overline{v})\right)^{\mathrm{T}},$$

whose components are given as follows:

$$\overline{\rho}_i^{\mathrm{B}}(\overline{v}) = \left[\frac{1}{2^{n-1}} \sum_{S \subseteq N \backslash i} (v_L(S \cup i) - v_L(S)), \frac{1}{2^{n-1}} \sum_{S \subseteq N \backslash i} (v_R(S \cup i) - v_R(S))\right] \quad (i = 1, 2, \ldots, n). \tag{3.57}$$

Example 3.12 Let us compute the interval-valued Banzhaf value of the interval-valued cooperative game $\overline{v}^0 \in \overline{G}^3$ given in Example 3.6.

It is obvious that the interval-valued cooperative game $\overline{v}^0 \in \overline{G}^3$ satisfies Eq. (3.17). Thus, according to Eq. (3.57), we can obtain

$$\begin{aligned}
\rho_{L1}^{\mathrm{B}}(\overline{v}^0) &= \frac{1}{2^2} \sum_{S \subseteq \{2,3\}} (\overline{v}_L^0(S \cup 1) - \overline{v}_L^0(S)) \\
&= \frac{1}{4}[(\overline{v}_L^0(1) - \overline{v}_L^0(\varnothing)) + (\overline{v}_L^0(1,2) - \overline{v}_L^0(2)) + (\overline{v}_L^0(1,3) - \overline{v}_L^0(3)) \\
&\quad + (\overline{v}_L^0(N') - \overline{v}_L^0(2,3))] \\
&= \frac{1}{4}[(0-0) + (2-1) + (1-2) + (2-2)] \\
&= 0,
\end{aligned}$$

$$\begin{aligned}
\rho_{R1}^{\mathrm{B}}(\overline{v}^0) &= \frac{1}{2^2}\sum_{S\subseteq\{2,3\}}(\overline{v}_R^0(S\cup 1)-\overline{v}_R^0(S))\\
&= \frac{1}{4}[(\overline{v}_R^0(1)-\overline{v}_R^0(\varnothing))+(\overline{v}_R^0(1,2)-\overline{v}_R^0(2))+(\overline{v}_R^0(1,3)-\overline{v}_R^0(3))\\
&\quad +(\overline{v}_R^0(N')-\overline{v}_R^0(2,3))]\\
&= \frac{1}{4}[(1-0)+(4-3)+(4-4)+(7-5)]\\
&= 1,\\
\rho_{L2}^{\mathrm{B}}(\overline{v}^0) &= \frac{1}{2^2}\sum_{S\subseteq\{1,3\}}(\overline{v}_L^0(S\cup 2)-\overline{v}_L^0(S))\\
&= \frac{1}{4}[(\overline{v}_L^0(2)-\overline{v}_L^0(\varnothing))+(\overline{v}_L^0(1,2)-\overline{v}_L^0(1))+(\overline{v}_L^0(2,3)-\overline{v}_L^0(3))\\
&\quad +(\overline{v}_L^0(N')-\overline{v}_L^0(1,3))]\\
&= \frac{1}{4}[(1-0)+(2-0)+(2-2)+(2-1)]\\
&= 1,\\
\rho_{R2}^{\mathrm{B}}(\overline{v}^0) &= \frac{1}{2^2}\sum_{S\subseteq\{1,3\}}(\overline{v}_R^0(S\cup 2)-\overline{v}_R^0(S))\\
&= \frac{1}{4}[(\overline{v}_R^0(2)-\overline{v}_R^0(\varnothing))+(\overline{v}_R^0(1,2)-\overline{v}_R^0(1))+(\overline{v}_R^0(2,3)-\overline{v}_R^0(3))\\
&\quad +(\overline{v}_R^0(N')-\overline{v}_R^0(1,3))]\\
&= \frac{1}{4}[(3-0)+(4-1)+(5-4)+(7-4)]\\
&= 2.5,\\
\rho_{L3}^{\mathrm{B}}(\overline{v}^0) &= \frac{1}{2^2}\sum_{S\subseteq\{1,2\}}(\overline{v}_L^0(S\cup 3)-\overline{v}_L^0(S))\\
&= \frac{1}{4}[(\overline{v}_L^0(3)-\overline{v}_L^0(\varnothing))+(\overline{v}_L^0(1,3)-\overline{v}_L^0(1))+(\overline{v}_L^0(2,3)-\overline{v}_L^0(2))\\
&\quad +(\overline{v}_L^0(N')-\overline{v}_L^0(1,2))]\\
&= \frac{1}{4}[(2-0)+(1-0)+(2-1)+(2-2)]\\
&= 1,
\end{aligned}$$

and

$$\rho_{R3}^{\mathrm{B}}(\overline{v}^0) = \frac{1}{2^2} \sum_{S \subseteq \{1,2\}} (\overline{v}_R^0(S \cup 3) - \overline{v}_R^0(S))$$

$$= \frac{1}{4}[(\overline{v}_R^0(3) - \overline{v}_R^0(\varnothing)) + (\overline{v}_R^0(1,3) - \overline{v}_R^0(1)) + (\overline{v}_R^0(2,3) - \overline{v}_R^0(2))]$$

$$+ (\overline{v}_R^0(N') - \overline{v}_R^0(1,2))]$$

$$= \frac{1}{4}[(4-0) + (4-1) + (5-3) + (7-4)]$$

$$= 3,$$

respectively. Hence, we obtain the interval-valued Banzhaf value of the interval-valued cooperative game $\overline{v}^0 \in \overline{G}^3$ as follows:

$$\overline{\boldsymbol{\rho}}^{\mathrm{B}}(\overline{v}^0) = ([0,1],[1,2.5],[1,3])^{\mathrm{T}},$$

which is remarkably different from the interval-valued Shapley value

$$\overline{\boldsymbol{\Phi}}^{\mathrm{SH}}(\overline{v}^0) = ([0,7/6],[1,8/3],[1,19/6])^{\mathrm{T}}$$

given in Example 3.6. Namely, we have

$$\overline{\rho}_i^{\mathrm{B}}(\overline{v}^0) \leq \overline{\Phi}_i^{\mathrm{SH}}(\overline{v}^0) \quad \left(i \in N' = \{1,2,3\}\right).$$

Furthermore, the interval-valued Banzhaf value $\overline{\boldsymbol{\rho}}^{\mathrm{B}}(\overline{v}^0)$ does not satisfy the efficiency, i.e.,

$$\sum_{i=1}^{3} \overline{\rho}_i^{\mathrm{B}}(\overline{v}^0) = [2,6.5] \neq \overline{v}^0\left(N'\right).$$

In the sequent, we discuss some useful and important properties of interval-valued Banzhaf values of interval-valued cooperative games.

Theorem 3.41 *(Existence and Uniqueness) For an arbitrary interval-valued cooperative game $\overline{v} \in \overline{G}^n$, if it satisfies Eq. (3.17), then there always exists a unique interval-valued Banzhaf value $\overline{\boldsymbol{\rho}}^{\mathrm{B}}(\overline{v})$, which is determined by Eq. (3.57).*

Proof According to Eq. (3.57), and combining with Definition 1.1, we can easily complete the proof of Theorem 3.1.

Theorem 3.42 *(Additivity) For any two interval-valued cooperative games $\overline{v} \in \overline{G}^n$ and $\overline{\nu} \in \overline{G}^n$, if they satisfy Eq. (3.17), then $\overline{\rho}_i^{\mathrm{B}}(\overline{v} + \overline{\nu}) = \overline{\rho}_i^{\mathrm{B}}(\overline{v}) + \rho_i^{\mathrm{B}}(\overline{\nu})$ $(i = 1,2,\ldots,n)$, i.e., $\overline{\boldsymbol{\rho}}^{\mathrm{B}}(\overline{v} + \overline{\nu}) = \overline{\boldsymbol{\rho}}^{\mathrm{B}}(\overline{v}) + \overline{\boldsymbol{\rho}}^{\mathrm{B}}(\overline{\nu})$.*

Proof According to Eq. (3.57) and Definition 1.1, we have

$$\overline{\rho}_i^{\mathrm{B}}(\overline{\upsilon}+\overline{\nu})=\left[\frac{1}{2^{n-1}}\sum_{S\subseteq N\setminus i}[(\upsilon_L(S\cup i)+\nu_L(S\cup i))-(\upsilon_L(S)+\nu_L(S))],\right.$$

$$\left.\frac{1}{2^{n-1}}\sum_{S\subseteq N\setminus i}[(\upsilon_R(S\cup i)+\nu_R(S\cup i))-(\upsilon_R(S)+\nu_R(S))]\right]$$

$$=\left[\frac{1}{2^{n-1}}\sum_{S\subseteq N\setminus i}(\upsilon_L(S\cup i)-\upsilon_L(S)),\frac{1}{2^{n-1}}\sum_{S\subseteq N\setminus i}(\upsilon_R(S\cup i)-\upsilon_R(S))\right]$$

$$+\left[\frac{1}{2^{n-1}}\sum_{S\subseteq N\setminus i}(\nu_L(S\cup i)-\nu_L(S)),\frac{1}{2^{n-1}}\sum_{S\subseteq N\setminus i}(\nu_R(S\cup i)-\nu_R(S))\right]$$

$$=\overline{\rho}_i^{\mathrm{B}}(\overline{\upsilon})+\overline{\rho}_i^{\mathrm{B}}(\overline{\nu}),$$

i.e.,

$$\overline{\rho}_i^{\mathrm{B}}(\overline{\upsilon}+\overline{\nu})=\overline{\rho}_i^{\mathrm{B}}(\overline{\upsilon})+\overline{\rho}_i^{\mathrm{B}}(\overline{\nu})\quad(i=1,2,\ldots,n).$$

Thus, we obtain

$$\overline{\boldsymbol{\rho}}^{\mathrm{B}}(\overline{\upsilon}+\overline{\nu})=\overline{\boldsymbol{\rho}}^{\mathrm{B}}(\overline{\upsilon})+\overline{\boldsymbol{\rho}}^{\mathrm{B}}(\overline{\nu}).$$

Therefore, we have completed the proof of Theorem 3.42.

Theorem 3.43 *(Symmetry) For any interval-valued cooperative game $\overline{\upsilon}\in\overline{G}^n$, if it satisfies Eq. (3.17), and players $i\in N$ and $k\in N$ $(i\neq k)$ are symmetric in the interval-valued cooperative game $\overline{\upsilon}$, then $\overline{\rho}_i^{\mathrm{B}}(\overline{\upsilon})=\overline{\rho}_k^{\mathrm{B}}(\overline{\upsilon})$.*

Proof According to Eq. (3.57), and combining with Definition 1.3 given in the previous Sect. 1.4.1, we can straightforwardly prove that the conclusion of Theorem 3.43 is valid.

Theorem 3.44 *(Anonymity) For any interval-valued cooperative game $\overline{\upsilon}\in\overline{G}^n$ and any permutation σ on the set N, if $\overline{\upsilon}$ satisfies Eq. (3.17), then $\overline{\rho}_{\sigma(i)}^{\mathrm{B}}(\overline{\upsilon}^{\sigma})=\overline{\rho}_i^{\mathrm{B}}(\overline{\upsilon})$ $(i=1,2,\ldots,n)$. Namely, $\overline{\boldsymbol{\rho}}^{\mathrm{B}}(\overline{\upsilon}^{\sigma})=\boldsymbol{\sigma}^{\#}(\overline{\boldsymbol{\rho}}^{\mathrm{B}}(\overline{\upsilon}))$.*

Proof According to Eq. (3.57), we can easily complete the proof of Theorem 3.44 (omitted).

Theorem 3.45 *(Null player) For any interval-valued cooperative game $\overline{\upsilon}\in\overline{G}^n$, if it satisfies Eq. (3.17), and $i\in N$ is a null player in the interval-valued cooperative game $\overline{\upsilon}$, then $\overline{\rho}_i^{\mathrm{B}}(\overline{\upsilon})=0$.*

Proof Due to the assumption that $i \in N$ is a null player in the interval-valued cooperative game $\overline{v} \in \overline{G}^n$. Then, according to Definition 1.4, we have

$$\overline{v}(S \cup i) = \overline{v}(S)$$

for any coalition $S \subseteq N \backslash i$, i.e.,

$$v_L(S \cup i) = v_L(S)$$

and

$$v_R(S \cup i) = v_R(S).$$

Therefore, according to Eq. (3.57), we directly have

$$\begin{aligned} \overline{\rho}_i^{\mathrm{B}}(\overline{v}) &= \left[\frac{1}{2^{n-1}} \sum_{S \subseteq N \backslash i} (v_L(S \cup i) - v_L(S)), \frac{1}{2^{n-1}} \sum_{S \subseteq N \backslash i} (v_R(S \cup i) - v_R(S))\right] \\ &= [0, 0] \\ &= 0. \end{aligned}$$

Namely, $\overline{\rho}_i^{\mathrm{B}}(\overline{v}) = 0$. Hereby, we have completed the proof of Theorem 3.45.

Theorem 3.46 *(Dummy player) For any interval-valued cooperative game* $\overline{v} \in \overline{G}^n$, *if it satisfies Eq. (3.17), and* $i \in N$ *is a dummy player in the interval-valued cooperative game* $\overline{v}$, *then* $\overline{\rho}_i^{\mathrm{B}}(\overline{v}) = \overline{v}(i)$.

Proof Due to the assumption that $i \in N$ is a dummy player in the interval-valued cooperative game $\overline{v} \in \overline{G}^n$, then according to Definition 1.5, we have

$$\overline{v}(S \cup i) = \overline{v}(S) + \overline{v}(i)$$

for any coalition $S \subseteq N \backslash i$, i.e.,

$$v_L(S \cup i) = v_L(S) + v_L(i)$$

and

$$v_R(S \cup i) = v_R(S) + v_R(i).$$

Hence, according to Eq. (3.57), we have

$$\begin{aligned}\overline{\rho}_i^{\mathrm{B}}(\overline{v}) &= \left[\frac{1}{2^{n-1}}\sum_{S\subseteq N\backslash i}(v_L(S\cup i)-v_L(S)),\frac{1}{2^{n-1}}\sum_{S\subseteq N\backslash i}(v_R(S\cup i)-v_R(S))\right]\\ &= \left[\frac{1}{2^{n-1}}\sum_{S\subseteq N\backslash i}v_L(i),\frac{1}{2^{n-1}}\sum_{S\subseteq N\backslash i}v_R(i)\right]\\ &= \left[\frac{1}{2^{n-1}}\times 2^{n-1}v_L(i),\frac{1}{2^{n-1}}\times 2^{n-1}v_R(i)\right]\\ &= [v_L(i),v_R(i)]\\ &= \overline{v}(i),\end{aligned}$$

i.e.,

$$\overline{\rho}_i^{\mathrm{B}}(\overline{v}) = \overline{v}(i).$$

Hereby, we have completed the proof of Theorem 3.46.

Theorem 3.47 (Invariance) For any interval-valued cooperative game $\overline{v}\in\overline{G}^n$ and its associated interval-valued cooperative game $\overline{\nu}\in\overline{G}^n$ given by Eq. (3.15), if they satisfy Eq. (3.17), then $\overline{\rho}_i^{\mathrm{B}}(\overline{\nu})=a\overline{\rho}_i^{\mathrm{B}}(\overline{v})+\overline{d}_i$ ($i=1,2,\ldots,n$), i.e., $\overline{\boldsymbol{\rho}}^{\mathrm{B}}(\overline{\nu})=a\overline{\boldsymbol{\rho}}^{\mathrm{B}}(\overline{v})+\overline{\boldsymbol{d}}$.

Proof According to Eqs. (3.20) and (3.15) and Definition 1.1, we have

$$\begin{aligned}\overline{\rho}_i^{\mathrm{B}}(\overline{\nu}) &= \left[\frac{1}{2^{n-1}}\sum_{S\subseteq N\backslash i}\left[\left(av_L(S\cup i)+\sum_{j\in S\cup i}d_{Lj}\right)-\left(av_L(S)+\sum_{j\in S}d_{Lj}\right)\right],\right.\\ &\qquad\left.\frac{1}{2^{n-1}}\sum_{S\subseteq N\backslash i}\left[\left(av_R(S\cup i)+\sum_{j\in S\cup i}d_{Rj}\right)-\left(av_R(S)+\sum_{j\in S}d_{Rj}\right)\right]\right]\\ &= a\left[\frac{1}{2^{n-1}}\sum_{S\subseteq N\backslash i}(v_L(S\cup i)-v_L(S)),\frac{1}{2^{n-1}}\sum_{S\subseteq N\backslash i}(v_R(S\cup i)-v_R(S))\right]\\ &\quad+\left[\frac{1}{2^{n-1}}\sum_{S\subseteq N\backslash i}d_{Li},\frac{1}{2^{n-1}}\sum_{S\subseteq N\backslash i}d_{Ri}\right]\\ &= a\overline{\rho}_i^{\mathrm{B}}(\overline{v})+[d_{Li},d_{Ri}]\\ &= a\overline{\rho}_i^{\mathrm{B}}(\overline{v})+\overline{d}_i,\end{aligned}$$

i.e.,

$$\overline{\rho}_i^{\mathrm{B}}(\overline{\nu}) = a\overline{\rho}_i^{\mathrm{B}}(\overline{v})+\overline{d}_i\quad(i=1,2,\ldots,n).$$

Hereby, we obtain

$$\bar{\rho}^{\mathrm{B}}(\bar{v}) = a\bar{\rho}^{\mathrm{B}}(\bar{v}) + \bar{d}.$$

Thus, we have completed the proof of Theorem 3.47.

Furthermore, interval-valued Banzhaf values of interval-valued cooperative games do not always satisfy the individual rationality and the efficiency. A specific illustrated example may be referred to Example 3.12 as above.

References

1. Branzei R, Branzei O, Alparslan Gök SZ, Tijs S. Cooperative interval games: a survey. Cent Eur J Oper Res. 2010;18:397–411.
2. Branzei R, Alparslan-Gök SZ, Branzei O. Cooperation games under interval uncertainty: on the convexity of the interval undominated cores. Cent Eur J Oper Res. 2011;19:523–32.
3. Li D-F. Fuzzy multiobjective many-person decision makings and games. Beijing: National Defense Industry Press; 2003 (in Chinese).
4. Owen G. Game theory. 2nd ed. New York: Academic Press; 1982.
5. Moore R. Methods and applications of interval analysis. Philadelphia: SIAM Studies in Applied Mathematics; 1979.
6. Branzei R, Tijs S, Alparslan-Gök SZ. How to handle interval solutions for cooperative interval games. Int J Uncertain Fuzziness Knowl Based Syst. 2010;18:123–32.
7. Han W-B, Sun H, Xu G-J. A new approach of cooperative interval games: the interval core and Shapley value revisited. Oper Res Lett. 2012;40:462–8.
8. Alparslan-Gök SZ, Branzei R, Tijs SH. Cores and stable sets for interval-valued games, vol. 1. Center for Economic Research, Tilburg University; 2008. p. 1–14
9. Alparslan-Gök SZ, Branzei O, Branzei R, Tijs S. Set-valued solution concepts using interval-type payoffs for interval games. J Math Econ. 2011;47:621–6.
10. van den Brink R. Null or nullifying players: the difference between the Shapley value and equal division solutions. J Econ Theory. 2007;136(1):767–75.
11. Casajus A, Huettner F. Null, nullifying, or dummifying players: the difference between the Shapley value, the equal division value, and the equal surplus division value. Econ Lett. 2014;122(2):167–9.
12. Driessen TSH, Funaki Y. Coincidence of and collinearity between game theoretic solutions. OR Spectrum. 1991;13:15–30.
13. Yager RR. OWA aggregation over a continuous interval argument with applications to decision making. IEEE Trans Syst Man Cybern Part B Cybern. 2004;34(5):1952–63.
14. Hukuhara M. Integration des applications measurables dont la valeur est un compact convexe. Funkcialaj Ekvacioj. 1967;10:205–23.
15. Stefanini L. A generalization of Hukuhara difference and division for interval and fuzzy arithmetic. Fuzzy Set Syst. 2010;161(11):1564–84.
16. Shapley LS. A value for n-person games. In: Kuhn A, Tucker A, editors. Contributions to the theory of games II, Annals of mathematical studies. Princeton: Princeton University Press; 1953. p. 307–17.
17. Radzik T, Driessen T. On a family of values for TU-games generalizing the Shapley value. Math Social Sci. 2013;65(2):105–11.
18. Branzei R, Dimitrov D, Tijs S. Shapley-like values for interval bankruptcy games. Econ Bull. 2003;3:1–8.

19. Alparslan Gök SZ, Branzei R, Tijs S. The interval Shapley value: an axiomatization. Cent Eur J Oper Res. 2010;18:131–40.
20. Alparslan-Gök SZ, Branzei R, Tijs S. Big boss interval games. Institute of Applied Mathematics, METU and Tilburg University, Center for Economic Research, The Netherlands, CentER DP 47 (preprint no. 103); 2008.
21. Li D-F. Linear programming approach to solve interval-valued matrix games. Omega. 2011;39 (6):655–66.
22. Li D-F. Linear programming models and methods of matrix games with payoffs of triangular fuzzy numbers. Heidelberg: Springer; 2016.
23. Driessen T. Cooperation games: solutions and application. Dordrecht: Kluwer Academic Publisher; 1988.
24. Joosten R. Dynamics, equilibria and values. PhD thesis. The Netherlands: Maastricht University; 1996.
25. Driessen T, Radzik T. A weighted pseudo-potential approach to values for TU-games. Int Trans Oper Res. 2002;9(3):303–20.
26. Driessen T, Radzik T. Extensions of Hart and Mas-Colell's consistency to efficient, linear, and symmetric values for TU-Games. In: ICM Millennium Lectures on Games, 2003. p. 129–146.
27. Casajus A, Huettner F. On a class of solidarity values. Eur J Oper Res. 2014;236(2):583–91.
28. Nowak AS, Radzik T. A solidarity value for n-person transferable utility games. Int J Game Theory. 1994;23(1):43–8.
29. Kamojo Y, Kongo T. Whose deletion does not affect your payoff? The difference between the Shapley value, the egalitarian value, the solidarity value, and the Banzhaf value. Eur J Oper Res. 2012;216(3):638–46.
30. Casajus A, Huettner F. Null players, solidarity, and the egalitarian Shapley values. J Math Econ. 2013;49:58–61.
31. Chameni Nembua C. Linear efficient and symmetric values for TU-games: sharing the joint gain of cooperation. Games Econ Behav. 2012;74:431–3.
32. Banzhaf JF. Weighted voting does not work: a mathematical analysis. Rutgers Law Rev. 1965;19:317–43.
33. Laruelle A, Valenciano F. Shapley-Shubik and Banzhaf indices revisited. Math Oper Res. 2001;26:89–104.
34. Owen G. Multilinear extensions and the Banzhaf value. Naval Res Logist Quart. 1975;22:741–50.
35. Alonso-Meijide JM, Carreras F, Fiestras-Janeiro MG. A comparative axiomatic characterization of the Banzhaf-Owen coalitional value. Decis Support Syst. 2007;43(3):701–12.
36. Pusillo L. Banzhaf like value for games with interval uncertainty. Czech Econ Rev. 2013;7:5–14.

Printed in the United States
By Bookmasters